Design and control of chemical process systems

Some McGraw-Hill books of related interest

Langley: Carbon Fibres in Engineering
Meleka: Electron-beam Welding
Bompas-Smith: Mechanical Survival
Redding: Intrinsic Safety
Handley: Industrial Safety Handbook

Design and control of chemical process systems

J. R. Borer

Engineering Specialist (Control Systems)
Monsanto Corporate Engineering Department (Europe)

London · New York · St Louis · San Francisco · Düsseldorf
Johannesburg · Kuala Lumpur · Mexico · Montreal · New Delhi
Panama · Paris · São Paulo · Singapore · Sydney · Toronto

Published by

McGRAW-HILL Book Company (UK) Limited
MAIDENHEAD . BERKSHIRE . ENGLAND

Library of Congress Cataloging in Publication Data

Borer, J   R
    Design and control of chemical process systems.

    Bibliography: p.
    1.   Chemical process control.   I. Title.
TP155.75.B67   660.2'81   74-10759
ISBN 0-07-084447-X

Copyright © 1974 McGraw-Hill Book Company (UK) Limited. All rights reserved. No part of this publication may be reproduced, stored in retrieval system, or transmitted, in any form or by any means, electronic, mechanical, photocopying recording, or otherwise, without the prior permission of McGraw-Hill Book Company (UK) Limited.

PRINTED AND BOUND IN GREAT BRITAIN

Contents

INTRODUCTION	1
CHAPTER 1 WHAT IS A SYSTEM	6
1.1 Level of resolution	6
1.2 System states and coupling	8
1.3 Degrees of freedom and control	12
1.4 Time dependent relationships	16
1.5 Transient behaviour	18
CHAPTER 2 SYSTEM DESCRIPTION AND DEFINITION	22
2.1 Mathematical representation of transient relationships	22
2.2 The response of a 'forced' system	29
2.3 Integral transforms	31
2.4 Modifying the response	35
CHAPTER 3 FROM THEORY TO PRACTICE	42
3.1 Transfer functions of simple systems	42
3.2 Transfer functions including control	45
3.3 Inertia in the process	50
3.4 Inertia in the control mechanism	55
3.5 Conclusion	57
CHAPTER 4 BASIC PRINCIPLES OF DESIGN OF SYSTEMS	58
4.1 Introduction to design	58
4.2 Analysis by 'frequency response'	60
4.3 Analysis by phase/gain plots	65
4.4 Design by the root locus method	69
CHAPTER 5 COMPENSATION BY CONTROL	77
5.1 'Proportional' control and error 'reset'	77
5.2 'Rate', 'predictive', or 'derivative' compensation	82
5.3 Other control actions	83

CHAPTER 6	INTRODUCTION TO MULTIVARIABLE SYSTEMS	87
	6.1 Models of systems	87
	6.2 System interaction	89
	6.3 Steady-state dimensions	92
	6.4 A simple example	95
CHAPTER 7	STATE SPACE THEORY	99
	7.1 Dynamic interaction	99
	7.2 Closed-loop stability	102
	7.3 Controllability	106
CHAPTER 8	MULTIVARIABLE COMPENSATION	109
	8.1 Decoupling and control	109
	8.2 Sampling techniques	112
	8.3 A multivariable algorithm	114
CHAPTER 9	CONTROL OF A REAL INTERACTIVE PROCESS	120
CHAPTER 10	SYSTEM IDENTIFICATION	128
	10.1 Identification of multivariable systems	128
	10.2 Other methods	134
CHAPTER 11	ON LINE COMPUTER CONTROL TODAY	140
APPENDIX I	MATRICES	143
APPENDIX II	SYSTEM CONTROLLABILITY	145
APPENDIX III	OBSERVABILITY OF PARAMETERS	147
BIBLIOGRAPHY		150
INDEX		152

Introduction

Because words are inadequate to define scientific facts, mathematics has become the language of science although words are the more natural means of expression. To many engineers and applied scientists, understanding and expression do not come easily in mathematical terms, and their excessive use can make practical interpretation more difficult. It is often not necessary, and never sufficient, that an engineer should follow the rigorous logic of mathematical derivation; it is always necessary, and often sufficient, that he understands the principles and relationships on which realities are based.

This book is devoted to introducing the concepts and some of the techniques available for the design of process plant systems. For any system to perform its intended functions it must be carefully designed, as a system.

The diagram shows the components of a simple space heating system. Steam heats the air by means of a finned tube heat exchanger and the temperature is controlled automatically by positioning a throttling valve in the steam line, condensed steam being discharged to a collection main above the heater and duct: a simple system and a very common arrangement. The plant manager is at a loss to understand why such a simple system is so much trouble. The control of the air temperature is hopeless, the steam traps cause water hammer and the heat exchanger unit has failed twice in six months with leaking welds. The appropriate plant engineers have investigated each problem, the utilities engineer the steam traps, the mechanical engineer the weld failures, and the instrument engineer the poor performance of the controls, but none of them can find the answer to his particular problems.

There are not three separate problems but one, and the cause is poor system design. There is nothing wrong with the construction of the heater or the steam traps or even the control mechanisms, but there has been inadequate appreciation of the behaviour of these as component parts of a *system*. How *do* they behave?

Before we can begin to answer that question we must define:

1. What is the system required to do?
2. Under what circumstances is it required to do it?

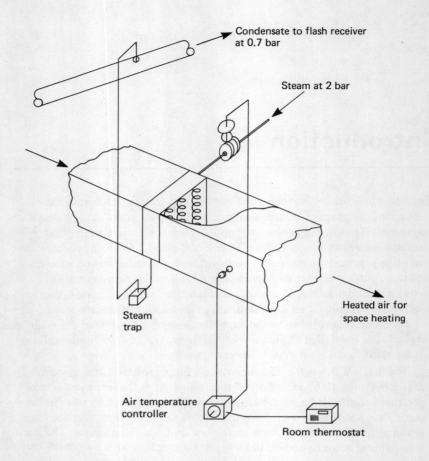

The system is intended to supply air at a steady rate and at such a temperature as will make up for the heat losses from the building. It is required to do this under a wide variety of environmental conditions, outside air temperatures of $-5\,°C$ to $+15\,°C$ and varying wind velocities. The heater has been designed to transfer the maximum quantity of heat required (at $-5\,°C$, etc.) from condensing steam, through the metal finned tubes, into the air. The steam trap is designed to discharge condensate while preventing the premature escape of steam. The control valve is designed to pass the quantity of steam required to be condensed at any time to achieve the air temperature demanded by the controller. Each of these components has been tested and performs its design functions satisfactorily, so why does the system not perform *its* function satisfactorily?

The quantity of heat transferred from steam to air depends on the surface available for the steam to condense on and the temperature gradient across it from steam side to air side. The steam condenses at a fixed temperature which depends on the steam pressure at the condensing face. The heater will undoubtedly have been overdesigned to provide a margin between its actual maximum heat transfer capability and the maximum duty required: even at maximum duty, therefore, the steam temperature will have to be lower than that for which the heater was designed in order to satisfy the heat transfer requirement. At low duties in spring and autumn the temperature difference across the full heat transfer surface might have to be about one-tenth the maximum. Now steam at approximately 2 bar condenses at 134°C. Under the dictates of the controller the valve progressively closes until the quantity of steam actually condensing equals that passing through the valve from the high pressure side to the lower pressure actually existing at that moment inside the heater. The pressure inside the heater can be calculated as follows:

Air temperature required	20 °C
steam temperature at 2 bar (Gauge)	134 °C
temperature difference steam/air	114 °C
required temp. diff. at one-tenth load	11.4 °C
hence required steam condensing temp.	(20 + 11.4) = 31.4 °C
corresponding pressure	approx. 45 mbar (abs)

This state of affairs will never be achieved in fact because long before the pressure has fallen to vacuum conditions the trap will have ceased to discharge condensed steam. In fact the trap will cease to work as soon as pressure in the heater is insufficient to lift the condensate up into the collection system overhead against the pressure (0.7 bar) in that main:

Height of condensate main above trap	approx. 3 m
Total pressure at trap	approx. 1 bar
condensing temperature at this pressure	120 °C
temperature difference steam/air	100 °C
corresponding to load reduction of $\frac{100}{114}$ =	0.88 of design rate

Condensed steam ceases to be discharged and remains inside the heater. The water is still at the condensing temperature at first and so heat is transferred into the air at first at exactly the same rate. The controller in a vain attempt to reduce this excessive rate causes the valve to close fully, shutting off steam altogether (unless it leaks) and eventually, as the collected condensate cools, the heat transfer rate reduces until the air temperature is lower than that required and the controller begins to open the valve. As the steam enters the heater it raises the pressure above 1 bar and the now cold condensate is discharged suddenly into the collection system by steam pressure. The heater is now empty of water, exposing the whole heat transfer surface once again to steam at high pressure and the heat transfer rate again greatly exceeds requirements.

The explanation for all three problems is now apparent. The controller and valve are quite unable to achieve the appropriate conditions in the heater at anything less than full load, if at all. The steam trap is prevented from discharging smoothly and continuously by the attempts of controller to achieve the impossible and slugs of cold condensate are discharged at intervals into the collection system causing water hammer, giving the impression that the trap is not capable of working correctly. Lastly the alternating and uneven temperature gradients within the heater set up stress conditions which, not surprisingly, lead to early fatigue failure at any welded joints.

The system does not appear to have been designed at all; each component has been designed against a definite specification, but no one has seen to it that these specifications reflect the actual behaviour of the component within the system. How could the system have been designed to work properly?

If the problem had been considered at an early enough stage it might have been possible to arrange for condensate to be discharged into a collection system at low level and at atmospheric pressure. If this were so and assuming that the height of the heater above this main exceeded approximately 10 m, a load reduction of about 10:1 would be possible with continuous condensate removal because the head of water which will build up in the line between the trap and the collection main now enables the pressure in the heater to be reduced below atmospheric pressure by the pressure equivalent of this head without restricting flow.

It is unlikely that 10 m elevation could be achieved and the condensate return system may have to be accepted as shown in the diagram. In this event we must make sure that sufficient pressure exists inside the heater at all times to force the condensate out. This can be achieved by replacing the throttling valve in the steam line by one in the condensate line between heater and steam trap. The condensate will now be discharged smoothly and continuously, but a level will be maintained in the heater tubes such that the combined heat transfer from condensing steam and cooling water is in equilibrium with that required to heat the air to the desired temperature. This method of operation, however, will only succeed provided the physical dimensions of the heater battery tubes are correctly designed. If the tubes are of excessive diameter and insufficient height, the quantity of condensed steam 'held up' per unit of heat transfer surface will be large (the heat transfer surface being proportional to diameter, the volume to diameter squared). Condensed steam will lose heat only slowly, and rapid changes of heat transfer rate will not be possible. In addition, while steam pressure will force condensate out at an acceptable rate to increase heat transfer by exposing more surface, the reduction of heat transfer by increase of water level will be dependent on the relationship of condensation rate to volume per unit of heat transfer surface. The tubes should therefore be long (high) and of narrow bore if good control of temperature is to be obtained.

It should be obvious already that the 'system design' cannot be isolated from either the design of the process itself or of the physical components which comprise the plant in which the process takes place. It should also be obvious that design of the system is not the same thing as the design of the process or the design of the plant items and it should not, as unfortunately it often is, be left to chance.

1. What is a 'system'

1.1 Level of resolution

'A word,' said Humpty Dumpty, 'means exactly what I want it to mean.' If he had in mind the word 'system,' his remark is not nearly so idiotic as it at first appears. If we are to consider the design of engineering systems as a discipline, it is essential that we start with a clear concept of what constitutes a 'system'. Perhaps the simplest definition, and one which provides a good starting point, is that a system is more than the sum of its component elements. A motor car comprises an engine, gear box, wheels, body, seats, etc., and the engine comprises pistons, crankshaft, carburettor, petrol pump, etc., but any enthusiastic motorist will agree that the total car is much more than just a collection of these 'bits'. What makes it so is the manner in which all these elements perform *together*, to achieve the overall objective, for each is only relevant as a component part of a system. The engine can be considered as a system, but it can also be considered in a wider context as a component part of the system which is the car.

The essential features of any system can be placed in one of two broad categories.

1. The *internal* component elements.
2. The relationships between these elements.

If one of an ensemble of elements is totally unrelated to any of the remainder, it cannot be part of the system since it is not affected by, nor does it have any affect on the system. A suitable definition for a system, then, is an ensemble or collection of two or more *related* elements. Within this definition we can draw the boundaries or limits of a system wherever we wish; for instance, we can consider the car as a system, including the driver, in order to assess how fast it goes, how well it handles, etc., or we may be concerned to assess the performance, torque, fuel consumption, etc., of the engine alone. We are concerned with the performance of the elements of the system under consideration not with their constructional details.

The system we chose to consider receives stimuli from its environment (I) and also influences its environment by producing responses (ϕ) to these stimuli. The car, for instance, pollutes the atmosphere which might be described as an unwanted response—its required response being a change of position.

It has already been noted that an element of the system under consideration may also be described as a system—the engine of the car, for instance. The car itself might be considered an element in a transportation *system*. There is no inconsistency here, for it can be seen from Fig. 1.1 that if the engine is an element a, of the system $a_1, a_2, a_3, \ldots, a_n$ which is the car, we can redefine the terms of reference by including all the other elements $a_2, a_3, \ldots, a_n$ in the environment set, so that $c_1, c_2, c_3, \ldots, c_n$ are elements of the engine (pistons, crankshaft, carburettor, etc.). We have chosen the 'level of resolution'

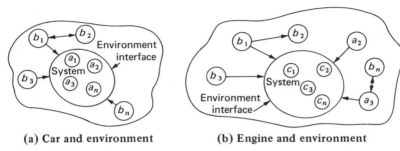

(a) Car and environment   (b) Engine and environment

Fig. 1.1.

which we wish to adopt in our study. We could not consider one piston within the engine as a system because it is a single indivisible element from a functional viewpoint and does not satisfy the definition of two or more related elements, but the carburettor could be considered a system in itself.

What then is the utility of the systems approach to engineering design? Engineers have not in the past required such an approach to the design of a car engine or indeed to the design of a car but it is certain that such a project as the moon 'shots' would not have been possible without a rigorous logical approach to the definition and analysis of systems. Established industries, such as the process industries, have developed from rudimentary beginnings methods of operation, construction, and design over many years, and it requires a fundamental change of direction both in management and executive functions to introduce the rigorous approach to design of complex systems which *is* systems engineering. Moreover the success or failure of a project in these industries is not so easily defined as it is with a moon-shot. Nevertheless, there is no doubt that a thorough understanding and application of the basic techniques of systems engineering will produce startling results in these industries and, indeed, is already beginning to do so in the larger and more advanced organizations.

The 'level of resolution' is established by the level of design in which the engineer is engaged; for instance, he may be concerned with the design of a refining column or with the interrelations of complete self-contained process plants forming part of a huge chemical complex. From a systems viewpoint the design of a car engine, for instance, is simple because the interrelations between the elements are to all intents and purposes instantaneous—there is no need to consider the storage of material or energy within the system. The manufacturing process of car engines on a production line, however, *is* a suitable case for applying systems engineering techniques. Even at the lowest levels of resolution within the process industries, design is concerned with the storage or 'hold up' of material and energy and relationships are invariably time dependent. It is this consideration more than any other which dictates the efficacy of systems engineering techniques to any design or analysis problem.

1.2 System states and coupling

The elements of any system are also elements of a larger set or ensemble which we might call the universal set, the remaining elements of which form the environment for the system under consideration. Thus, the remainder of the car is included in the environment of the engine and woe betide the manufacturer who forgets this. The car is part of the environment not the whole of it; the cold outside air which causes the carburettor to ice up is also part of the environment. For the sake of realism we would include in our universal set *all* elements which have any influence on (relationship with) any of the elements of the system under consideration. Since all the elements of the system must by definition be related, any element of the environment influences *all* the elements of the system and we can conveniently consider two sets of elements in adopting the systems approach.

A system which is influenced by its environment is an 'open system' and must exist in equilibrium with its environment. In process systems this implies the laws of conservation of matter and energy, and, except in so far as elements within the system can store either, the rate of gain to the system and the rate of loss from the system of material and energy must be equal. A 'closed system' is one which is entirely isolated from its environment, and such a system must exist in a state of *internal* equilibrium; perhaps the only true closed system is the universe itself. An open system is always part of a larger system which includes the environment of the system under consideration. By establishing the 'level of resolution' we draw in the boundaries of that section of the 'universal set' which we intend to make self-regulating. By doing so we ensure that the system, while performing its intended functions, remains in equilibrium with its environment, for unless it does so it cannot continue to perform its intended functions for any length of time. The behaviour of a system is defined in terms of the 'states' of its internal elements. Each 'system element' will perform some function of storing or transferring either material or energy and its 'state' will be a parameter which reflects this function (some elements may have an energy state and a material state).

The outputs of an open system are a function of some or all of its internal states. Any change in the inputs will affect some of the internal states: if the affect of any change of input is 'coupled' through the interactions of internal states, so that it produces a change in output which tends to restore the equilibrium between system and environment, the system is said to display self regulation. For such a system there will be a unique set of internal states for any given set of inputs. A system which is 'uncoupled' with respect to any input will be driven to some extreme limiting state by the smallest change in that input. Such a system is of no use; it may be modified, however, by the addition of controls to 'couple' the input to the output and make it self-regulating. Four simple systems are shown in Fig. 1.2 in order to illustrate the points made so far. The internal states of system I are h the head in the vessel and F the flowrate of fluid through the fixed restriction. The output is a simple function of F—it is equal to it. The single input influences h which is

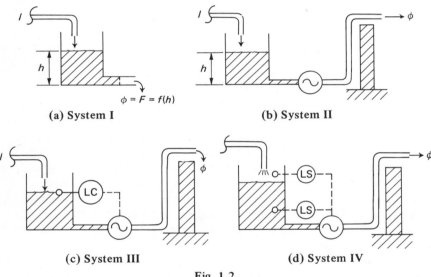

(a) System I (b) System II

(c) System III (d) System IV

Fig. 1.2.

related to F or ϕ in such a manner that when input I increases, output ϕ responds by increasing also, so restoring equilibrium in respect of material gain and loss. This system, which is 'coupled' in such a manner that it displays 'self-regulation', is stable and has a unique internal state h for each and every value I of the input. It also demonstrates the effect of capacity within a system; a sudden increase in input will result in h *starting* to increase and with it ϕ. It will be some time, however, before h and ϕ attain new 'steady-state' values and restore equilibrium, and meanwhile the system stores both material and energy within the vessel as represented by the increase in h. The rate at which h increases is dependent on the extent of the mismatch between input I

and output ϕ, but, since any increase in h reduces this mismatch, the rate of increase of h becomes less and less and the approach of the system states to their new steady values is as shown in Fig. 1.3.

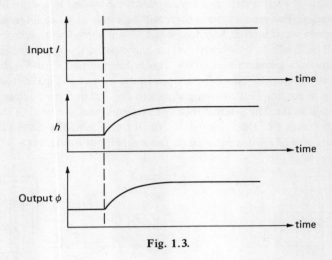

Fig. 1.3.

System II (see Fig. 1.2) is similar to system I but fluid is now pumped out over a wall. The pump is driven by a synchronous electric motor at constant speed and its rate of transfer of fluid ϕ is constant. A change of input still affects h as before, but the output ϕ is uncoupled—there is no internal relationship between h and ϕ. Unless I happens to exactly equal ϕ, h will either increase until limited by the overflow, or decrease until the vessel is empty. The system is 'divergent' and hence unstable, its states tending to diverge from a steady condition when it is stimulated, instead of converging onto another steady condition as does a coupled, self-regulating system.

In system III the addition of a measuring element M and controlling element C establishes a relationship between h and ϕ by varying the pump speed as h varies. The measuring and controlling 'elements' are not system elements since their function is not the storage or transfer of energy or material, but the establishing of a relationship: we shall see later that the controlling and measuring elements must often be considered as system elements when they do store or transfer relatively significant amounts of energy.

In system IV the pump is switched on at high level to pump out the tank and off at low level to allow it to fill up again. When h is between high and low levels the system behaves like system II. The controls added this time, however, cause a steep change of ϕ at discrete values (high and low limits) of h, thus introducing a relationship between h and ϕ. This system does not have a unique set of internal states for every input and is said to be 'discretely coupled'.

In order to study systems and their elements in terms of their behaviour it is necessary to establish a 'relating function' between inputs (stimuli) and outputs (responses). If this function is to be complete it must take into account the 'transient' behaviour of the internal system states during the period when the system is out of equilibrium with its environment, that is the period of time between steady-states. The existence of capacity within any system does not merely delay the response so that it occurs later in time than the stimulus, it also changes its distribution in time as can be seen from Fig. 1.3. The stimulus occurs suddenly, and the response or output *starts* immediately, but is only completed after an interval of time proportional to the *capacitance* of the vessel (the rate of change of h per unit rate of change of input). Pure delay between stimulus and response occurs in many systems when changes in the state of one element must travel (along pipe, conveyor belts, etc.) to reach another element.

The states of a system are the 'dimensions' of that system (not the dimensions of the physical elements but of the relationships of the system).

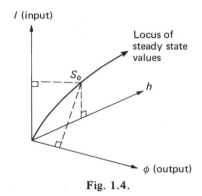

Fig. 1.4.

Figure 1.4 represents the system I in terms of a three-dimensional 'space'. The analogy to volume measured in three *linear* dimensions is readily understood, but the abstract conception of an n-dimensional 'space' cannot be represented on paper. Since there is a unique *steady-state* value of both h and ϕ for each value of I, the locus of such values, as I changes, will be along a line as shown, any particular set of corresponding values I, h, and ϕ being given by dropping perpendiculars onto the I, h, and ϕ axes. No such locus exists for system II. It must be realized that the locus is of *steady-state* values only. The locus of values of I, h, and ϕ corresponding to the transient behaviour of the system I following the stimulus of an instantaneous change of I is shown in Fig. 1.5.

11

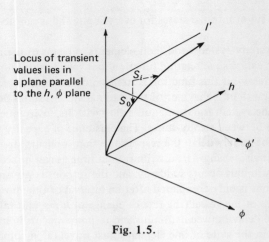

Fig. 1.5.

Note that a sudden increase of I moves the 'operating point S_0 to position S_i'. Since the change is instantaneous no *immediate* change in h and therefore none in ϕ occurs and S moves parallel to the I axis to represent this. Subsequently the value of I remains constant, the values h and ϕ moving along a projection of the steady-state locus in a plane parallel to the h, ϕ plane and perpendicular to the I axis until the new steady state is attained.

1.3 Degrees of freedom and control

This system has three dimensions I, ϕ, and h when considered in terms only of steady states; there are only two relationships resulting from physical laws to be satisfied:

1. $h = \int_0^T (\phi - I)\, dt$ Head equals the integral of the mismatch of input flow rate and output flow rate with respect to time.
2. $\phi = f(h) = R\sqrt{h}$ Outflow rate is a *constant* function of head. R being constant.

Since there are three dimensions and only two constraining relationships the system has one 'degree of freedom'. The locus of all possible operating points lies in one dimension that is, in graphical terms, along a line. If the system had three physical relationships to satisfy, the locus of operating points would be a single point, as there are no degrees of freedom and only one value is possible.

The number of constraints or relationships in *any* system *cannot* exceed the number of independent variables. Relationships, however, may connect two variables only or may interrelate several of the system states.

If we now redefine the system, this time taking into account its transient behaviour, we find that there are not three but four variables

$$h, \phi, I, \text{ and } \frac{dh}{dt}$$

Without control there are still only two relationships defining physical laws which the system is constrained to satisfy:

1. $\phi = f(h)$
2. $h = \int_0^\tau (\phi - I)\, dt$

dh/dt which was previously assumed to be zero to give steady-state relationships is now given by differentiating (2) thus

$$\frac{dh}{dt} = (\phi - I)$$

Since there are now $(4 - 2) = 2$ degrees of freedom, the loci of all possible operating conditions (steady state and transient) lie in a plane containing the I axis and the steady-state locus as shown in Fig. 1.6 since graphically two dimensions constitute a plane area:

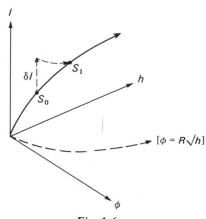

Fig. 1.6.

The locus of transient operating conditions following a step, or instantaneous, change of input I is shown here. The head h and outflow ϕ change exponentially with passage of time towards their new steady-state values, but maintain the relationship $\phi = R\sqrt{h}$. Since I does not vary further after its initial instant change of value the locus must lie in a plane (only two variables ϕ and h) parallel to the ϕ, h plane and at right-angles to the I axis as shown.

The addition of a control mechanism as shown in Fig. 1.7 completely changes the system.

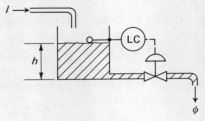

Fig. 1.7.

There are now four states: I, ϕ, h, and R. R is the resistance of the valve and is now variable. There are three steady-state relationships:

1. $\phi = R\sqrt{h}$ as before except that R is now variable
2. $h = \int_0^\tau (\phi - I)\, dt$
3. $R = f(H - h)$ H is head desired.

The function f relating R and $(H - h)$ can be such as to make $(H - h) = 0$ in steady state. Thus $h = H$ and only one value of h and one corresponding value of R are possible in steady state. However, there is $(4 - 3) = 1$ degree of freedom and the locus of steady-state operating points lies on a line (one dimension of freedom) which is perpendicular to the h axis as shown in Fig. 1.8.

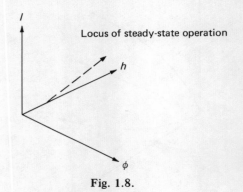

Fig. 1.8.

Redefining this new system (with control) taking into account the transient behaviour we find that two new states and one new relationship are added:

States: $I, \phi, h, dh/dt, R, dR/dt$.
Relationships:

1. $\phi = R\sqrt{h}$
2. $h = \int_0^\tau (\phi - I)\, dt$
3. $R = f(H - h)$

dh/dt and dR/dt are obtained by differentiating 1 and 3 respectively.

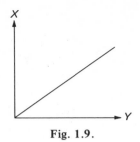

Fig. 1.9.

In defining the states in this way we are ignoring the rate of change of ϕ as an independent variable and assuming that ϕ follows h in fixed relationship instantly. This is realistic in practice since any delays between ϕ and h are so small as to be insignificant compared with the delays which occur between changes of I and corresponding changes of either h or R.

There are two degrees of freedom of transient behaviour in addition to the one degree of freedom in steady state. In the special case where all system relationships are linear as shown in Fig. 1.9 (and they can often be made approximately linear) a single degree or dimension of transient freedom means that transition from one steady state to another takes place in a single direction with passage of time, as can be seen in Fig. 1.6. The approach of any single-state system to its new steady value is exponential in respect of time, becoming slower and slower as the new state is approached (Fig. 1.10). In such a case, the system state will be nearer to the new steady value after each increment of time, *never* further away. If, however, there are two or more dimensions of transient freedom the locus of states during the time interval between steady values is defined by two or more coordinates and can change direction (even if all system relationships are linear). Hence the value of the state *may* be further away from the new steady value at successive intervals in time, the approach to the new steady value, with respect to time, being as shown in Fig. 1.11.

Provided there exists only one degree of transient freedom the transition

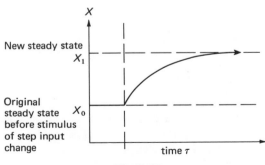

Fig. 1.10.

15

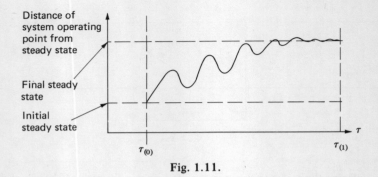

Fig. 1.11.

from one steady state to another will not be oscillatory even if system relationships are not linear. It requires a *reversal* of the *direction* of the locus of transient operating points to produce oscillation. If the non-linearity is such that there is a reversal of slope as shown in Fig. 1.12, then there is no longer a unique value of Y for any value of X; two possible values of Y exist for the

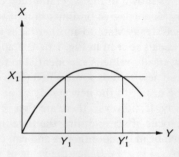

Fig. 1.12.

one value X. A common example of such a non-linear relationship exists in the pressure/flowrate characteristic of centrifugal compressors and is associated with the phenomenon of surge which is oscillatory in nature. The practical solution is to constrain, or limit, the system to operate only over a portion of the relationship.

1.4 Time dependent relationships

All but the simplest elements of any process plant behave in a fashion which is time dependent to some extent. Hitherto this has been largely ignored in design of process plant systems: in operating a process it cannot be ignored. For this reason transient behaviour of plant elements has been regarded as something to be avoided and controls applied to suppress it. Process plant has been designed to operate continuously in 'steady state', disturbances of any sort being avoided as far as possible. Most difficult to control are those plants designed for discontinuous operation, for during start up and shut down transient behaviour cannot be avoided. The extent to which process plant can be 'automated', however, depends entirely on the ability to define in precise terms the relationships of stimuli to response within the system. The term 'automated' is used here to describe a plant which, while continuing to carry out the function for which it is designed, is capable of responding in a stable manner to any normal stimulus. Thus an 'automated' plant requires no human operative for normal operation.

Time dependent (transient) relationships are relationships of growth and decay: any quantity which neither grows nor decays is not time dependent. Living organisms grow by successive division of cells starting originally with one single cell, but growth of populations of individuals takes place by a different mechanism, individuals being indivisible units. The 'population' of a vessel consists of the individual molecules of the fluid contents. The mechanism of growth of the contents of the vessel and of a human population, is exactly the same—by addition of new individuals rather than division of those already there.

If we trace back in time the growth of any population, we must (as with the single cell origin of a living organism) arrive at a time when only one unit existed—an Adam and Eve situation. The preliminary stages of the growth of a population is therefore discrete, but as the population grows the generations become mixed and the addition of new individuals distributed in time so that growth becomes a continuous process for practical purposes. The rate at which individuals are added is directly proportional to the total population at any time—the rate of increase is proportional to the 'state' of the population.

Similarly, the rate of decay of any population is the result of the removal of individuals—death. If the population (or system) considered is in equilibrium with its environment, the additions and removals will balance and the popu-

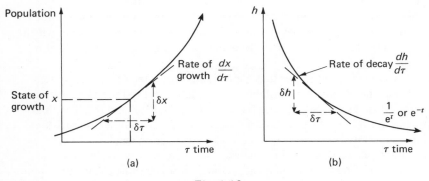

Fig. 1.13.

lation achieves some steady-state value in balance with outside influences such as decrease, predators, etc.

The 'population' of the vessel is expressed as h and the removal rate ϕ is proportional to h. When the addition rate I and ϕ are equal the population or total contents is 'steady'—h does not alter, but when there is a mismatch between I and ϕ, h must change, either growing or decaying.

The mathematical function of which the rate of growth is proportional to state of the parameter is the exponential e and therefore expressions of growth are always time functions of e and expressions of decay are inverse time functions ($1/e^{xt}$ or e^{-xt}). In the vessel the initial rate of growth decays as the 'population' increases—the inverse function of e (Fig. 1.13).

1.5 Transient behaviour

The transient behaviour of the internal states of any system, when stimulated by some change of input always comprises first derivative (rate of change or velocity) terms and often second and even higher derivatives of the variable with respect to time. The state equations, which define the total relationships of input, internal states, and output are therefore differential equations embodying both transient and steady-state relationships.

The behaviour of a system, then, during the time that its inputs and outputs of energy and material are out of balance following a stimulus, depends on the growth and decay relationships *and delays* within the system and the nature of the stimuli applied.

The simple system used as an example earlier—the fluid filled vessel—displays the typical exponential response associated with a *single* storage element—the vessel. Complex systems usually comprise many storage elements both for energy and for material: sometimes energy and material storage elements are combined, sometimes separate. In addition, the transfers of either, within the system, may be subject to pure 'transport' delays. We must now consider how these factors interact to give an infinite variety of transient behaviour, which sometimes results in sustained maloperation and, hence, ruined product and even severe damage to plant, in a system which is not actually unstable in the sense that we have given this word. A system which exhibits unacceptable transient behaviour is sometimes, wrongly, called unstable.

It was mentioned earlier that controls are very often applied to system elements, not to make an unstable system stable, but to modify a steady state within that system. Such a case is shown in Fig. 1.14(a) and (b). The fluid within the vessel (a) will assume a unique level for any given combination of input and output values. If it is required that the level remains constant, a measuring element, control mechanism, and final element (valve) must be added as shown in (b). In an earlier example it was assumed that the response of the controls was instantaneous, so that these elements could be considered

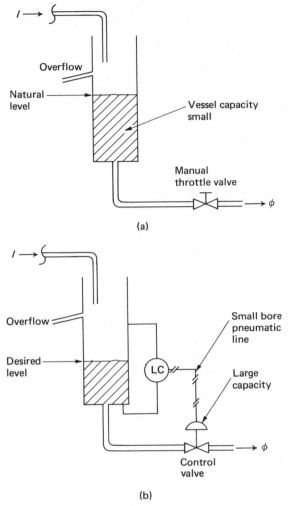

Fig. 1.14.

as establishing a relationship while not themselves constituting 'plant elements'. If, however, the control elements store energy, they become plant elements. The control in this example is pneumatically operated, compressed air entering or leaving the large diaphragm 'motor' of the control valve to position the air as dictated by the control mechanism. The diaphragm chamber constitutes a capacity for storing air pressure, just as the vessel constitutes a capacity, for storing not only liquid, but also the potential energy of the liquid. Let us suppose that the pipe along which the air must travel to and from the diaphragm chamber is of very small diameter and offers considerable resistance to the passage of air. If now an abrupt or 'step' change in the rate of inflow occurs due to opening the valve, the level in the vessel will *start* to rise immediately; the measuring element will detect this and air will *start* to flow into the diaphragm chamber. It will, however, be some time before the pressure in the chamber has built up sufficiently to cause the valve to open to such a position that outflow again equals inflow. By this time the level in the vessel will be considerably higher than desired, and the control mechanism (the objective of which is to maintain a constant level) will continue to cause the valve to open so that outflow exceeds inflow. The level in the vessel will start to fall and the control valve will eventually cease to open and the level will fall to that desired. Nevertheless, the level will continue to fall, as outflow now exceeds inflow and the control mechanism will start to close the valve. Again it will be some time before sufficient air can escape from the diaphragm chamber so that the level in the vessel ceases to fall, by which time it will be considerably lower than intended. The control mechanism will now cause the valve to close further so that the inflow will once again exceed the outflow and the level rise. For the same reason that, as it fell, the level overshot the desired value it will now overshoot as it rises:

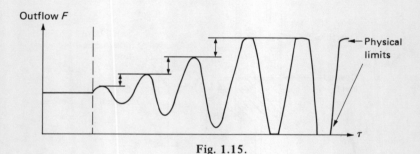

Fig. 1.15.

The level may now exceed the first peak, in which case the whole process will be repeated in successive cycles, each peak exceeding the last, until eventually the system alternates between the overflowing condition and empty (see Fig. 1.15). The system is unstable, for although the input and output are 'coupled', the desired state h can no more be attained than if it were not so. Such a system is both divergent *and oscillatory* in its behaviour.

Whether successive overshoot peaks are greater or smaller than previous ones will be determined by the 'dimensions' of the system. If successive peaks diminish, the oscillations will eventually die out leaving the system in a new steady state. Such a system is stable, though its response is oscillatory. It is such oscillatory behaviour which is sometimes, incorrectly, referred to as instability, for, from a practical point of view, such transient behaviour may be unacceptable.

For a system to display oscillatory transient response, there must exist within the system more than one element of energy storage. Either the storage elements will be physically separate or energy will be stored in more than one form, often within the same physical element. Energy is transfered from one storage element to another and back again or is converted from one form to another and back again as the case may be. In this example energy is stored as potential energy—the head of fluid—and as kinetic energy—the outflow velocity—and it is the delayed cyclical interchange of these two energy forms which gives rise to the oscillation of material within the vessel. Another example of an oscillatory system (or elements) is a pendulum in which the reversible interchange of potential energy and kinetic energy of the mass give rise to oscillatory displacement.

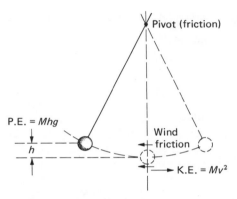

Fig. 1.16.

If a pendulum is set swinging its oscillations will gradually die away because some of the total energy is converted by friction to heat which cannot then be reconverted and is therefore lost to the system: eventually the pendulum must come to rest in a 'minimum energy' position. In its normal state the pendulum is a 'closed' system (Fig. 1.16). By setting it swinging we demonstrate its natural transient response, but there is no continuous input and output of energy; thus, after the transient response has decayed, it resumes its original state of equilibrium. Not so the vessel system; it receives a continuous input of potential energy and returns to the environment continuous outputs of velocity energy and heat due to friction. This system is normally 'open' and can be either stable or unstable.

The vessel with level control can behave in an oscillatory fashion and may be an element (or sub-system) of a larger system, say, a refining column. Other elements of the larger system may have as their input the output of this sub-system, so that they may be 'forced' or stimulated by an oscillatory input. It is necessary therefore that we express oscillatory functions mathematically and this will be the subject of the next chapter.

2. System description and definition

2.1 Mathematical representation of transient relationships

Before we can define the relationship of two quantities (the input and output of a system) which are changing with the passage of time, we must appreciate that two quite distinct types of change are possible. The first type encompasses any steady change of variable which leads eventually to a 'permanent' or 'steady-state' change. For example, the changing level in a vessel without control mechanism, following a change in input rate, leads to the establishment of a new level as has been seen. The second type of change (oscillatory) is of a special nature and can be defined as cyclical motion or change, about an *unchanging* mean value; the variable concerned is constantly changing, but in a fashion that does not lead to any permanent alteration. The behaviour of any changing quantity is described as transient (varying with time), but these two types, although often superimposed, must not be confused—they are in every sense independent. For this reason oscillatory quantities are mathematically described as 'imaginary' numbers (prefixed by $j\sqrt{(-1)}$) and steady-state changes as 'real' numbers.

The justification for this use of j can be demonstrated by taking a mathematical liberty and squaring the infinite series

$$+1 -1 +1 -1 + \ldots$$

which is clearly an oscillatory sequence the average value over any even number of terms being zero. Because it is an infinite series having no beginning and no end the squared value can be obtained either by multiplying like terms

$$+1 -1 +1 -1 + \cdots$$

multiplied by

$$+1 -1 +1 -1 + \cdots$$

giving

$$+1 +1 +1 +1 + \cdots$$

multiplied by -1 throughout gives

$$\cdots -1 -1 -1 -1 - \cdots$$

If a variable j is defined such that

$$[(j)(\cdots +1 -1 +1 -1 \cdots)]^2 = (\cdots -1 -1 -1 \cdots)$$

then

$$(j)^2 \times (\cdots +1 -1 +1 -1 \cdots)^2 = -i(\cdots +1 -1 +1 \cdots)^2$$

hence $j \equiv \sqrt{-1}$. Now any changing quantity must, as has been shown, be expressed as a function of the exponential e which can itself be expressed as an infinite series—an irrational number. Therefore multiplying the infinite series e^{xt} by j accurately expresses an oscillating quantity, and for convenience we write this as

$$j\, e^{xt} \quad \text{where } j = \sqrt{-1} \qquad (2.1)$$

Steady-state relationships in a system are those which apply when none of the inputs or outputs is disturbed. Such relationships are customarily expressed as equalities or equations in algebraic form. For a dynamic system the relationships comprise both steady-state relations and also relations of rates of change, and these can be expressed in the form of differential equations. It has been seen that changing quantities are expressed by terms of the form

$$c\, e^{xt} \quad \text{(where } c \text{ is a scale factor)}$$

so we would expect the solution to a differential equation to comprise such terms.

Consider the simple differential equation

$$\frac{dy}{dt} - ky = 0 \quad \text{(where } k \text{ is a constant)} \qquad (2.2)$$

This can be written

$$\frac{d}{dt}(y) - k(y) = 0$$

or $py - ky = 0$ where $p = d/dt$ and is called an 'operator'. Hence $y(p - k) = 0$.

The expression $(d/dt)(y)$ or py means 'the rate of change, with respect to time, of the variable y'. The equation $y(p - k) = 0$ must include the solution of the simple algebraic equation

$$(p - k) = 0 \quad \text{(called the 'characteristic equation')}$$

i.e. $p = k$ whatever the value of y may be; hence $py = ky$ or 'the rate of change of the variable y is proportional to the state of y'. This, of course, is the definition of an exponential function of time and so the only possible value of y which will satisfy is

$$y = c\, e^{pt} = c\, e^{kt} \quad (c \text{ is a scale constant}) \tag{2.3}$$

This result can also be obtained by integrating the equation thus

$$\frac{dy}{dt} + ky = 0$$

multiplying by dt and dividing by y throughout

$$\frac{1}{y}.\,dy + k.\,dt = 0$$

hence integrating with respect to y and again with respect to t the left-hand term being a function of y only and the right-hand term a function of t only

$$(\log_e y) + kt = A \quad \text{where } A \text{ is a constant of integration.} \tag{2.4}$$

Taking anti-logs: $y\, e^{kt} = c$ where c is another constant hence $y = c\, e^{-kt}$. It can be seen that the constant c is derived from the constant of integration.

Now consider the equation for the motion of a pendulum

$$\frac{d^2y}{dt^2} + k_1 \frac{dy}{dt} + k_2 y = 0 \tag{2.5}$$

which can be rewritten

$$p^2 y + k_1 p y + k_2 y = 0$$

or

$$y(p^2 + k_1 p + k_2) = 0 \tag{2.6}$$

If e^{-pt} is a solution (that is a value of y which satisfies the equation) then by definition y can be replaced by it in the above equation, hence

$$e^{-pt}(p^2 + k_1 p + k_2) = 0$$

and a requirement for $e^{-pt} = y$ to be a solution since e^{-pt} can never be zero is

$$(p^2 + k_1 p + k_2) = 0 \quad \text{(the characteristic equation).}$$

Solutions to the characteristic equation are given by

$$p = \frac{-k_1 \pm \sqrt{(k_1^2 - 4k_2)}}{2} \quad (2.7)$$

which gives two solutions or roots, and thus the general solution to the original equation is

$C_1 e^{pt} + C_2 e^{p_2 t}$ where p_1 and p_2 are the two roots of the characteristic equation.

Since the equation contains a second-order differential term it could be integrated twice to obtain the solution and this would result in two constants of integration (C_1 and C_2). It can therefore be seen that a third-order differential equation solution will contain three terms, a fourth order four, and so on.

A simple nth order differential equation with constant coefficients is solved by forming the characteristic equation, solving it and then finding the constants $C_1, C_2, \ldots, C_n$ by inserting particular known values into the result.

So far all values have been real or non-oscillatory quantities and the solution $C_1 e^{p_1 t} + C_2 e^{p_2 t} + \cdots + C_n e^{p_n t}$ is a sum of real growth and decay terms. But how are oscillatory quantities to be expressed?

In the solution of the second-order characteristic equation

$$p = \frac{-k_1 \pm \sqrt{(k_1^2 - 4k_2)}}{2} \quad (2.8)$$

it is possible that $4k_2 > k_1^2$ giving a term which is the root of negative number. Multiplying under the root sign by -1 makes $(k_1^2 - 4k_2)$ positive and the root can be found. This is equivalent to multiplying $\sqrt{(k_1^2 - 4k_2)}$ by $\sqrt{-1}$ or j so the result becomes

$$p = \frac{-k_1 \pm j\sqrt{(-k_1^2 + 4k_2)}}{2}$$

which is known as a complex number

$$p = \frac{k_1}{2} \pm \frac{1}{2j} \{\sqrt{(-k_1^2 + 4k_2)}\}$$

The two alternatives

$$p = \frac{k_1}{2} + \frac{1}{2j} \{\sqrt{(-k_1^2 + 4k_2)}\}$$

and

$$p = \frac{k_1}{2} - \frac{1}{2j} \{\sqrt{(-k_1^2 + 4k_2)}\}$$

are known as a conjugate pair of complex numbers. The general solution to a second-order equation is

$$y = C_1 e^{p_1 t} + C_2 e^{p_2 t} \qquad (2.9)$$

where p_1 and p_2 and also C_1 and C_2 may be complex conjugate pairs.

If k_1 is zero the result is wholly oscillatory, i.e.,

$$\left. \begin{array}{l} p_1 = +\dfrac{1}{2j}(\sqrt{4k_2}) \\[6pt] p_2 = -\dfrac{1}{2j}(\sqrt{4k_2}) \end{array} \right\} \qquad (2.10)$$

and

and the solution to the equation

$$\frac{d^2 y}{dt^2} + k_2 y = 0 \quad \text{where } k_1 \text{ has been set} = 0 \text{ in the general form}$$

is seen to be

$$y = C_1 e^{\frac{\sqrt{4k_2}}{2j} t} + C_2 e^{-\frac{\sqrt{4k_2}}{2j} t}$$

or

$$y = C_2 e^{j\sqrt{k_2} t} + C_1 e^{-j\sqrt{k_2} t} \qquad (2.11)$$

Now if $y = 0$ at $t = 0$ (that is the pendulum is at centre position to begin with; an arbitrary decision which decides at what point $t = 0$) then $e^{j\sqrt{k_2} t} = e^0 = 1$ and $0 = C_2 + C_1$ and since C_2 and C_1 are a conjugate pair

$$C_2 = \frac{1}{2j} \quad \text{and} \quad C_1 = -\frac{1}{2j} \quad \text{satisfies this requirement.}$$

Hence a solution to $d^2 y/dt^2 + k_2 = 0$ is

SINE WAVE →
$$\frac{1}{2j}(e^{j\sqrt{k_2} t} - e^{-j\sqrt{k_2} t}) \qquad (2.12)$$

which is the expression for a sine wave whose frequency is given by $\omega = \sqrt{k_2}$ so that

$$\frac{d^2 y}{dt^2} + \omega^2 y = 0 \quad \text{where} \quad \frac{d}{dt} = y\omega \text{ or } -y\omega$$

is an alternative expression for oscillatory change derived from the more general expression

$$\frac{d^2 y}{dt^2} + k_1 \frac{dy}{dt} + \omega^2 y = 0 \qquad (2.13)$$

Applying a second set of initial conditions which are equally valid, i.e., $y = 1$ when $t = 0$ (which means that the pendulum is at an extreme position when $t = 0$) then

$$1 = C_1 + C_2$$

and $C_1 = C_2 = \frac{1}{2}$ satisfies this identity; hence another solution to $d^2y/dt^2 + \omega^2 = 0$ is

COSINE WAVE → $\frac{1}{2}(e^{j\omega t} + e^{-j\omega t})$

which is in fact the exponential form of a cosine function. Now ($y = 1$ when $t = 0$) and ($y = 0$ when $t = 0$) are initial conditions which, when considered with respect to periodic time ωt, are 90° (or $\pi/2$ radians) out of phase. A general solution to the purely transient equation $d^2y/dt^2 + \omega^2 = 0$ is then

$$y = A \sin \omega t + B \cos \omega t \tag{2.14}$$

in which the relative values of A and B are determined by an arbitrary decision as to when $t = 0$.

Putting back the $k_1(dy/dt)$ term into the equation it can be seen that instead of terms

$$e^{+j\sqrt{k_2}t} \quad \text{and} \quad e^{-j\sqrt{k_2}t}$$

the solution to the general equation

$$\frac{d^2y}{dt^2} + k_1 \frac{dy}{dt} + \omega^2 y = 0 \tag{2.15}$$

comprises terms

$$e^{(k_1 t + j\sqrt{k_2}t)} \quad \text{and} \quad e^{(k_1 t - j\sqrt{k_2}t)}$$

These can be re-written

$$e^{k_1 t} \times e^{\pm j\sqrt{k_2}t}$$

and the general solution to the general second-order equation is

$$y = e^{k_1 t}(A \sin \underline{\omega} t + B \cos \underline{\omega} t) \tag{2.16}$$

where $\underline{\omega}$ is the 'damped' frequency, not the 'natural' frequency (ω). The solution is seen to be complex; there is a steady-state change superimposed on the oscillation. In the case of the pendulum, $e^{k_1 t}$ represents friction of the pivots and air. The oscillatory motion gradually dies away until a stable state is reached when there is only one value possible, minimum potential energy together with minimum velocity energy. The inclusion of the $k_1(dy/dt)$ term therefore implies frictional damping. It is easier to see if the expression (2.16)

27

is put into another form, thus,

Replacing A and B by $\sin \psi$ and $\cos \psi$

where $A/B = \tan \psi$: $y = e^{-k_1 t}(\sin \psi \cos \omega t + \cos \psi \sin \omega t)$ (2.17)

Hence
$$y = e^{-k_1 t} \sin(\omega t + \psi) \tag{2.18}$$

where ψ is an arbitrary phase angle determined by when $t = 0$, from which it can be seen that the sine wave, $\sin(\omega t + \psi)$, dies away at an exponential rate $e^{-k_1 t}$. The sine wave represents the reversible energy exchange from kinetic to potential energy and the term $e^{-k_1 t}$ the decay of total energy due to irreversible change, i.e., friction.

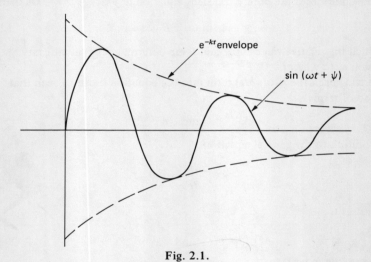

Fig. 2.1.

Graphically the damped or decaying oscillation is represented as in Fig. 2.1. The general expression for damped oscillatory motion is

$$\frac{d^2 y}{dt^2} + 2\zeta\omega \frac{dy}{dt} + \omega^2 y = 0 \quad (\zeta \text{ is called the damping factor}) \tag{2.19}$$

to which the solution is of the form

$$y = e^{-2\zeta\omega t} \sin(\omega t + \psi) \tag{2.20}$$

So far we have only considered the natural behaviour of a normally closed system, such as the free pendulum, which is momentarily interfered with. Such a system must eventually come to rest after the transient response to the momentary disturbance has died away. What of the behaviour of an open system, such as the vessel system of earlier examples: if the normally steady input is momentarily disturbed, returning afterwards to its original state then this system behaves in the same way as the pendulum (if it is oscillatory). If the input is suddenly changed to a new constant value oscillatory transient response results as before, but the system eventually comes to rest in a new steady state, reflecting the change in input. A stimulus, however, does not usually take place instantly and may consist of a steadily changing input or, indeed, the oscillatory output of another system.

Whatever the form of the stimulus, after an interval during which the transient response dies away, the response will be of the same form, delayed in time and changed in amplitude, but otherwise the same as will now be shown.

2.2 The response of a 'forced' system

The equation for a forced second-order system can be written

$$\frac{d^2y}{dt^2} + 2\zeta\omega\frac{dy}{dt} + \omega^2 y = f(t) \quad \text{(where } f(t) \text{ is the stimulus or 'forcing function')}$$

or

$$P^2 y + 2\zeta\omega P y + \omega^2 y = f(t)$$

or

$$y(P^2 + 2\zeta\omega P + \omega^2) = f(t) \tag{2.21}$$

and the response

$$y = \frac{1}{P^2 + 2\zeta\omega P + \omega^2} \cdot f(t) \tag{2.22}$$

This general case of the equation expresses both oscillatory and steady-state components of change and the operator P is in consequence complex. However, if the stimulus $f(t)$ is pure oscillation—($x \sin \omega' t$)—there can be no steady-state component of change and $P = j\omega'$.

Hence, if

$$f(t) = x \sin \omega' t$$

(ω' is the frequency of the forcing function which is not in general the same as the 'natural' frequency ω)

$$y = \frac{x}{P^2 + 2\zeta\omega P + \omega^2} \sin \omega t$$

can be rewritten

$$y = \frac{x}{-(\omega')^2 + 2\zeta\omega j\omega' + \omega^2} \sin \omega' t \tag{2.23}$$

which can be re-expressed

$$\left. \begin{array}{l} \text{output } y = \left[\dfrac{1}{(\omega^2+\omega'^2)} + \dfrac{1}{j(2\zeta\omega\,.\,\omega')}\right] x \sin \omega' t \\[2ex] \text{or } \quad y = \left[\dfrac{1}{(\omega+\omega')(\omega-\omega')} + \dfrac{1}{j(2\zeta\omega\,.\,\omega')}\right] x \sin \omega' t \end{array} \right\} \quad (2.24)$$

Now the term in the square brackets is a complex quantity which can be expressed as magnitude and phase hence the output y which after an initial period must also be sinusoidal differs from the input forcing function in magnitude and phase only.

It can readily be seen that as the frequency of the 'forcing' function approaches the natural or free frequency of the element or system being stimulated, the first term in the squared brackets tends towards infinity. Thus, the response tends towards an infinitely large sinusoid as this condition is approached. This phenomenon, known as 'resonance', occurs in electronic tuned capacitative inductive circuits (known as 'acceptor' circuits) and sometimes in process systems where a reciprocating gas compressor is coupled to pipe-work systems in which high velocities and capacity of lines and vessels give the same tuned effect. In an electronic circuit the capacity stores potential energy while the inductance stores velocity energy; in the compressor system cyclical interchange of velocity (or kinetic energy) and pressure (or potential energy) takes place, often with dire results.

The special nature of pure oscillatory stimulation of a system is that it produces a continuous transient response which contains no steady-state component and thus 'discovers' the transient behaviour of the system stimulated. If the operator P is replaced by $j\omega'$ in the equation

$$y = \frac{1}{(P^2 + 2\zeta\omega P + \omega^2)} x \sin \omega' t \qquad (2.25)$$

to give

$$y = \frac{1}{(\omega')^2 + 2j\zeta\omega\,.\,\omega' + \omega^2} x \sin \omega' t \qquad (2.26)$$

then

$$\boxed{\frac{y}{x \sin \omega' t} = \frac{1}{(\omega')^2 + 2j\zeta\omega\,.\,\omega' + \omega^2}} \qquad (2.27)$$

which is called the 'transfer function' and relates input to output in terms of transient behaviour only. If the stimulus is pure oscillation the only behaviour possible is transient. There is in fact a transient component of this transient behaviour initially, since a system with mass and inertia cannot immediately change from rest into oscillatory motion, but takes time to be 'pulled' into synchronous oscillation by the stimulus.

2.3 Integral transforms

All time varying quantities which might constitute stimuli can be decomposed into pure oscillations. This fact is quite familiar to most people, for a pure oscillation of air pressure within the audible range constitutes a 'pure note', while all musical sounds, indeed, all sounds, are composed of pure tones added together in various proportions. The radio engineer is well aware that sounds can be decomposed or synthesized at will if a wide enough 'spectrum' of frequencies or pure oscillations is available. We are also aware from this analogy that decomposition and synthesis is not restricted to a continuous function (sound), but can be achieved for a finite stimulus which has definite starting and finishing times.

If somehow, we can 'decompose' the input or stimulus into its component 'pure' frequencies and write a mathematical expression for this, then that expression multiplied by the transfer function (expressed as a complex function) will give the output, expressed in the same way as the input.

Any *cyclic* quantity can be decomposed into it's component frequencies, each of which, being a sinusoid, is composed of terms $e^{nj\omega t}$ where n is an integer between $-\infty$ and ∞. Thus the decomposed function can be expressed

$$f(\omega t) = \sum_{n=-\infty}^{\infty} C_n e^{nj\omega t} \qquad (2.28)$$

where for any value of n C_n is a weighting constant. Each C_n is obtained by multiplying both sides of (2.28) by

$$e^{-nj\omega t} \qquad (2.29)$$

and integrating over the period. Thus

$$C_n = \frac{1}{T} \int_0^T f(\omega t) \cdot e^{-nj\omega t} \, d(\omega t) \qquad (2.30)$$

We now have a mathematical expression for a decomposed input function, provided this is cyclic. In general, however, input functions will not be cyclic; we must find an equivalent expression for non-cyclic functions.

Can non-periodic functions be decomposed into elemental functions $e^{jn\omega t}$? To say that a function is non-periodic is really to say it has a period of $T = \infty$. Since $\omega = 2\pi/T$, ω will become smaller as T becomes larger. Designating $\Delta\omega$ as the value of ω when T is very large, since $T = 2\pi/\Delta\omega$, then

$$C_n = \frac{\Delta\omega}{2\pi} \int_{-\pi/\Delta\omega}^{\pi/\Delta\omega} f(t) e^{-jn\Delta\omega t} \, dt \qquad (2.31)$$

Substituting for C_n in (2.28):

$$f(t) = \sum_{n=-\infty}^{\infty} \left[\left\{ \frac{\Delta\omega}{2\pi} \int_{-\pi/\Delta\omega}^{\pi/\Delta\omega} f(t) e^{-jn\Delta\omega t} dt \right\} e^{jn\Delta\omega t} \right]$$

$$= \sum_{n=-\infty}^{\infty} \left[\left\{ \frac{1}{2\pi} \int_{-\pi/\Delta\omega}^{\pi/\Delta\omega} f(t) e^{-jn\Delta\omega t} dt \right\} e^{jn\Delta\omega t} \right] \Delta\omega$$

taking limits

$$\sum_{n=-\infty}^{\infty} \{\quad\} \Delta\omega \text{ tends to } \int_{n=-\infty}^{\infty} \{\quad\} d\omega$$

as $n\Delta\omega$ tends to ω and $\Delta\omega$ tends to 0. Hence

$$f(t) = \int_{-\infty}^{\infty} \frac{1}{2\pi} \left\{ \int_{-\infty}^{\infty} f(t) \cdot e^{-j\omega t} dt \right\} e^{j\omega t} d\omega$$

and defining a new function $F(\omega) = \int_{-\infty}^{\infty} f(t) e^{-j\omega t} dt$

$$f(t) = \frac{1}{2\pi} \int_{-\infty}^{\infty} F(\omega) e^{j\omega t} d\omega$$

$F(\omega)$ is called the 'Fourier transform' of the time varying function $f(t)$. It is the function which describes the amplitude of each frequency or pure oscillation which occurs in $f(t)$ when this is decomposed into an infinite number of such pure oscillations. Graphically represented it is a plot of amplitude against frequency as shown in three diagrams of stereotyped functions of time in Fig. 2.2.

Since the Fourier transform decomposes into pure oscillation it can only be used where the function decomposed is itself periodic or of the type shown above in which the parameter returns to the same value after some period of time. We cannot use the Fourier transform to decompose the 'step' function shown in Fig. 2.3 because the response to such a function will contain steady state as well as oscillatory transients. If we wish to 'discover' both steady state and oscillatory transient behaviour we must decompose into elements which combine real change with imaginary, that is, elements of the form $e^{(\sigma+j\omega)t}$ not $e^{j\omega t}$. The transform becomes

$$F_{(\sigma-j\omega)} = \int_{-(\sigma+j\omega)}^{(\sigma+j\omega)} f(t) e^{-(\sigma+j\omega)t} dt \qquad (2.32)$$

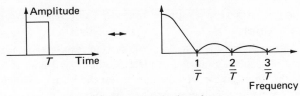
(a) A rectangular impulse.

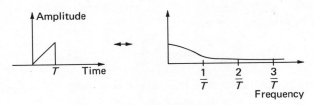
(b) A 'sawtooth' impulse.

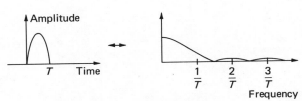
(c) A half-sine impulse.

Fourier transforms   Impulse time functions

Fig. 2.2.

or

$$F(s) = \int_{-(\sigma - j\omega)}^{(\sigma + j\omega)} f(t)\, e^{-\sigma} + e^{-j\omega} \cdot dt \qquad (2.33)$$

where s is used to denote the complex operator.

This is the Laplace transform and, while it is beyond the scope of this book to elaborate further on the theory of integral transforms, the following proper-

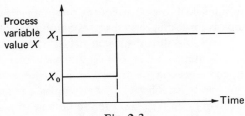
Fig. 2.3.

ties of the Laplace, which are not properties of the Fourier transform should be understood:

1. The final value of the time function—the system response—is obtained by letting *t tend to* ∞. This is equivalent to letting *s tend to* 0 since $e^{-0} = e^0 = 1$. Thus, the final value or steady state, after the initial transient behaviour has decayed, is neither growing or decaying in 'real' terms, though there may remain a continuous oscillatory component.

2. Similarly, the initial value can be obtained by letting *t tend towards* 0 which is equivalent to *s* tending to ∞ since

$$e^{-\infty} = \frac{1}{e^{\infty}} = \frac{1}{\infty} = 0$$

3. A necessary condition for the integration to be possible is that the function decays, however slowly, since the integral is the area under the function curve which is infinite unless this condition is met as shown in Fig. 2.4.

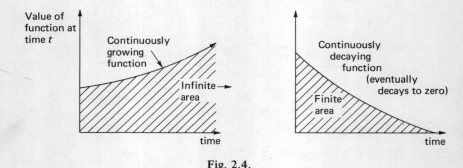

Fig. 2.4.

The Laplace operator $s\,(\sigma + j\omega)$ is a very useful tool, for if it is made to replace $j\omega$ in the transfer function as previously defined, the transfer function becomes the complete relating function of response to stimulus so that the relationship

$$\phi(s) = (RF)(s) \times I(s)$$

gives the transient response to *any* stimulus, and not simply to the special case of a sinusoidal stimulus.

The response to sinusoidal forcing—frequency response—is very important because it forms a basis for testing systems to establish the relating function. By superimposing a sinusoidal disturbance on a system, in the absence of other disturbances, recording the ratio of magnitude and phase displacement of output to input, and repeating the process for a full spectrum of frequencies, the transfer function can be established. If the complex operator *s* is substituted thereafter for ω or $j\omega$ in the expression so obtained the response for any stimulus can be calculated.

2.4 Modifying the response

If the system response can be found, it can be used to estimate the output which will result from any given input. The input which, if applied through the manipulable variable, will annihilate this response can then be calculated. If these calculations can be made in a time which is insignificant with respect to the system response, it will be possible to prevent any significant error developing. Another method of control is to measure the output, compare it to a desired value and manipulate the input to correct for any difference or error. The first method described is known as feedforward control and has the advantage that it prevents errors developing whereas the second method, known as feedback control, corrects only for an error which has already occurred. However, feedback control does not presuppose a detailed knowledge of the system response, or entail calculation of output responses and is therefore much simpler to design. The disadvantage of feedback control lies in

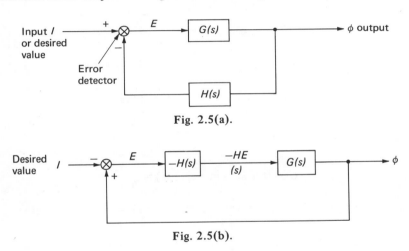

Fig. 2.5(a).

Fig. 2.5(b).

the fact that a complete closed 'loop' is formed; input–system–output–controller–input, which makes possible oscillatory response and instability. Both feedforward and feedback control can be used together on occasions, the former based on approximate calculation, reducing, though not obviating completely the output error, while the latter corrects for the smaller error which does in fact develop.

A simple feedback control loop is illustrated in 'block diagram' form in Fig. 2.5(a). (The block diagram is a very useful representation.) This system has only one input and one output, and disturbances can only occur either at the input or the output. Hence the system can, in this case, be represented by one block $G(s)$; the control mechanism is similarly represented by a single block $H(s)$. Even relatively simple systems have to be broken down into several blocks to permit consideration of other inputs and outputs. Control mechanisms must also on occasions be subdivided for consideration of the measuring elements, the error detector, the final control element, etc. An alternative form of control mechanism is shown in Fig. 2.5(b) in which the error and not the measured value signal is amplified. The principle of the block diagram is that the output of any block is the input multiplied by the function represented by the block. If the output of one block is the input of another in series with it, it is only necessary to multiply by the transfer function to obtain the input/output relationship for both.

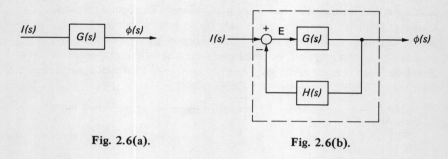

Fig. 2.6(a).   Fig. 2.6(b).

Consider first the system without control (Fig. 2.6(a))

$$\phi(s) = I(s) \times G(s)$$

or

$$G(s) \text{ (the system response)} = \frac{\phi(s)}{I(s)}$$

Then the response of the system with control added is given by Fig. 2.6(b).

$$G(s) \text{ now} = \frac{\phi(s)}{E(s)} \text{ as it's input is now } E(s) \quad (2.34)$$

and

$$E(s) = I(s) - H(s)\phi(s) \quad (2.35)$$

Eliminating $E(s)$ to obtain an expression for $\phi(s)/I(s)$ from (2.34) and (2.35)

$$G(s) = \frac{\phi(s)}{I(s) - H(s).\phi(s)} \quad (2.36)$$

hence

$$\phi(s) = G(s)[(I(s) - H(s).\phi(s))]$$

or

$$\phi(s)[(1 + G(s).H(s))] = G(s).I(s)$$

Therefore

$$\frac{\phi(s)}{I(s)} = \frac{G(s)}{1 + G(s).H(s)} \tag{2.37}$$

The advantage of the block diagram representation is immediately apparent, for the system with control can now be treated as one block with the transfer function

$$\frac{G(s)}{1 + G(s).H(s)} \text{ *}$$

Consider a system in which a disturbance enters neither at the manipulated input, nor at the output, but at some intermediate point. Divide the system into two parts $G_1(s)$ and $G_2(s)$ about the point of entry of the disturbance $D(s)$ as in Fig. 2.7. In order to establish the relationship $\phi(s)/D(s)$, $I(s)$ must be eliminated. There is no reason why instruments cannot be calibrated with the

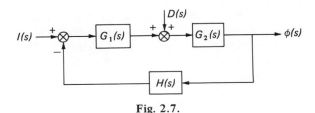

Fig. 2.7.

steady-state values as arbitrary zeros and since no disturbances enter the system through the input $I(s)$, and since its value on the chosen scales is zero, $I(s)$ can be eliminated and the block diagram redrawn (Fig. 2.8), hence the transfer function for this input/output relationship is

$$\frac{\phi(s)}{D(s)} = \frac{G_2(s)}{1 + G_1(s).G_2(s).H(s)} \tag{2.38}$$

The basic flexibility of the block diagram is evident here and the reader may note that considerable analysis and design is possible without considering the nature of the blocks or subdivisions of the system at all, thus greatly simplifying the design of more complex systems. He should also note at this point that so far only one input/output pair at a time has been considered; the coupling effects are difficult though not impossible to analyse. Later in this book it will be shown how a multi-input/output system can be considered in its entirety.

* The alternative control mechanism has a transfer function

$$\frac{G(s)H(s)}{1 + G(s)H(s)}$$

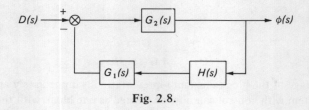

Fig. 2.8.

In chapter 1 irreversible energy conversion leading to real growth or decay of a quantity was represented in the form of a first-order differential equation; reversible energy conversion leading to oscillatory response was seen to be a feature of second-order differential expressions. The existence of a first-order term in a second-order expression was shown to imply damping of the oscillatory response and to lead to real or irreversibly changed states of energy. The first- and second-order time differentials are, of course, known as velocity and acceleration, and find common expression in everyday experience. Unfortunately there are no common place examples of higher-order equations, a fact which is illustrated by the lack of words in our language to describe higher-order differentials (like 'velocity' and 'acceleration'). However, quite simple systems may contain higher-order terms in their descriptions. There is no difficulty in extending the principles outlined already to higher-order expressions. The roots of the second-order equation

$$p^2 + k_1 p + k_2 = 0$$

can be expressed in factored form

$$(p + a)(p + b) = 0$$

where $k_1 = a + b = 2\zeta\omega$ and $k_2 = ab = \omega^2$. Similarly the roots of an nth-order equation can be expressed in factored form

$$(p + a)(p + b)(p + c) \cdots (p + n) = 0 \qquad (2.39)$$

It has been shown that the characteristic equation forms the denominator of the transfer function so that the transfer function of an nth-order system can be expressed as

$$\frac{K}{(p + a)(p + b) \cdots (p + n)} \qquad (2.40)$$

where K is a constant which represents the attenuation or magnification produced by the system, which depends on the range over which the input and output functions respectively are measured. For instance, the quantity of steam supplied to a hot water heater is measured over a range of 0–100 kg/h, and the temperature of the water passing through this heater (at a constant rate) is measured over the range 40–50 °C, a change of 10 kg/h in the steam flowrate causing a 10 °C change in the water temperature. In this case a 10 per cent input change causes 100 per cent output change and $K = 10$.

The 'complex' characteristic function which forms the denominator of the transfer function $(p + a)(p + b) \cdots (p + n)$ determines the transient behaviour of the system, while K is a measure of the sensitivity of the output to change of input.

Now consider the simple second-order system with the block diagram shown in Fig. 2.6.

Let

$$G(s) = \frac{K}{(s+a)(s+b)}$$

the transfer function

$$\frac{\phi(s)}{I(s)} = \frac{G(s)}{1 + G(s) \cdot H} \qquad (2.41)$$

where H represents the feedback loop gain, assuming the response of the control mechanism to be instantaneous. Hence

$$\frac{\phi(s)}{I(s)} = \frac{\dfrac{K}{(s+a)(s+b)}}{1 + \dfrac{KH}{(s+a)(s+b)}}$$

$$= \frac{K}{(s+a)(s+b) + KH}$$

$$= \frac{K}{s^2 + (a+b)s + (ab + KH)} \qquad (2.42)$$

The characteristic equation of this controlled system (in which $G(s)$ is the system to be controlled and H the control mechanism in a feedback loop) is seen to be a second-order function which includes the sensitivity constant K. The roots of the characteristic equation

$$s^2 + (a+b)s + (ab + KH) = 0 \qquad (2.43)$$

are

$$s = \frac{-(a+b) \pm \sqrt{\{(a+b)^2 - (4ab + 4KH)\}}}{2}$$

and it can be seen that the value of KH will determine whether the roots are wholly real as when

$$(a+b)^2 > 4(ab + KH) \qquad (2.44)$$

or complex, introducing oscillatory response, when

$$(a+b)^2 < 4(ab+KH) \tag{2.45}$$

A stable system is one in which all terms e^{st} are decay terms, i.e., s has negative real values only. The imaginary parts of s, since they must always in practice occur as conjugate pairs, if they occur at all, are purely transitory and therefore can, at worst (in the undamped case), lead to continuous oscillation, not runaway (unstable) response.

Again, it can be seen that all values of s will always be negative provided a, b, and KH are always positive, but s *can* become positive, leading to unstable response, if either a, b, or KH are negative (*note:* s is not *necessarily* + in such a case). Now a and b are the roots of the characteristic equation for the system without feedback control, $G(s)$, so if a or b are negative one of the values of s in the system characteristic equation is + and the system without control is therefore unstable. However, if KH is a suitable value, the system *with* feedback control is not unstable. If a and b are both + then all values of s in the system (without control) are—and the system is stable even without control; in this case the only way in which the system can become unstable with feedback control is if KH is negative which can occur if the phase of the output with respect to the input is such as to reverse the intended feedback from negative to positive. This can occur with high frequency or fast changing inputs.

A knowledge of the sensitivity or magnitude ratio of input to output together with a knowledge of the phase relationship of input to output will enable one to determine whether a system, with or without feedback control, is stable or unstable to given inputs or disturbances. The simplest feedback control mechanisms consist of a variable gain, allowing H to be set at various values greater or less than unity, so that the sensitivity KH is variable and can be adjusted to compensate for a system which, without control, is unstable (has a + root s) or to avoid making an otherwise stable system unstable to 'fast' inputs. A knowledge of how the roots of the controlled system's characteristic equation change with variations of H the control mechanism gain will indicate what values of H can be tolerated without making any system unstable.

From all that has been said it will be seen that although data, in the form of magnitude ratio and phase relationship, can be obtained by testing a system, the quantity of data required to cover an adequate range of frequencies and an adequate range of control gain settings is excessive. More sophisticated techniques using pulse inputs which enable the response to a whole range of frequencies to be obtained in one test have been developed. Most recently the development of statistical or stochastic techniques have made it possible to test a system using low amplitude random noise or an approximation to it which enables tests to be carried out on running systems.

In the field of calculation for design, if the system transfer functions can be calculated even approximately it is possible to plot the locus of the roots of the

CE of a controlled system as they change with variation of controller gain by the 'root locus' method which will be briefly described later.

The basic principles of one form of system control, feedback control, have been outlined: feedback control mechanisms are often more complex than a simple variable gain amplifier/attentuator and can be designed to compensate for undesirable features of the response of a system, as will be seen later.

3. From theory to practice

3.1 Transfer functions of simple systems

It has been shown that the most useful description of the dynamics of elements (or blocks) of a system is the transfer function and that this can be obtained by writing the differential equation for the system. The relationships between the differential equation, the transfer function, and the physical dimensions and behaviour of real systems will now be demonstrated.

Consider a simple tank (Fig. 3.1) with liquid inflow of I litres per second and outflow through the valve of ϕ litres per second. The rate of outflow depends on the resistance R of the valve and the head h of liquid in the tank.

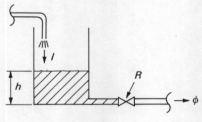

Fig. 3.1.

Starting with an empty tank the inflow will all go initially to increase the head since, until some head is established, no outflow can occur at all. Gradually, however, as the head increases the outflow will increase until such a head is reached eventually as will support an outflow rate ϕ exactly equal to the inflow I. This is the steady state of the system with constant inflow I.

The dimensions of the system can be defined as the capacitance of the tank, C, and the resistance of the valve, R. C is defined as the rate of increase in the measured output variable h per unit increase in the input variable I. R is defined as the head h required to support an outflow of one unit.

It can be seen that the steady-state head established depends on the dimension R and on the constant inflow rate I; a different head will be established for each combination of these variables.

The transient behaviour of the variable h is described by the first-order differential equation

$$\frac{dh}{dt} = \frac{1}{C}(I - \phi) \tag{3.1}$$

For the moment it will be assumed that the relationship of h to ϕ is linear ($R\phi = h$) so that R is a simple constant. This is not normally so as flow is more often turbulent than viscous.

Since $\phi = h/R$ the equation in 'operator' form is

$$sh = \frac{1}{C}\left(I - \frac{h}{R}\right) \tag{3.2}$$

or

$$sh + \frac{h}{CR} = \frac{I}{C} \tag{3.3}$$

Now if I commences at a constant rate and instantly (the definition of a step input), inflow = 1 at and after $t = 0$. The transfer function obtained from the characteristic function is

$$\frac{h(s)}{I(s)} = \frac{1/C}{\left(s + \frac{1}{CR}\right)} \tag{3.4}$$

The dimensions of C are

$$\frac{\text{length cubed}}{\text{length}} = \text{length}^2$$

The dimensions of R are

$$\frac{\text{length}}{\text{length cubed/time}} = \frac{\text{length} \times \text{time}}{\text{length cubed}} = \frac{\text{time}}{\text{length}^2}$$

hence the dimensions of CR are

$$\frac{\text{length}^2 \text{ time}}{\text{length}^2} = \text{time}$$

CR is known as the 'time constant' of the process, T. The transfer function can therefore be written

$$\frac{1/C}{\left(s + \frac{1}{T}\right)}$$

The, by now familiar, transient solution to the first-order differential equation (4.3) is

$$h = k e^{-t/T} \tag{3.5}$$

The particular or steady-state solution obtained by setting $s = 0$ is

$$\frac{h}{T} = \frac{I}{C} \quad \text{or} \quad \frac{h}{CR} = \frac{I}{C} \tag{3.6}$$

hence $h = I . R$.

The complete solution is therefore

$$h = k e^{-t/T} + IR \tag{3.7}$$

and the constant k is given by substituting the known condition that the tank is empty, $(h = 0)$ at $t = 0$, hence

$$0 = k e^0 + IR$$
$$= k + IR$$
$$k = -IR$$

and

$$h = IR - IR \, e^{-t/T} \quad \text{or} \quad IR(1 - e^{-t/T}) \tag{3.8}$$

Since the second (transient) term is decaying, it will eventually disappear leaving the steady-state solution

$$h = IR \quad \text{or} \quad I = \frac{h}{R} = \phi$$

that is, inflow equals outflow at a head $h = IR$.

The system can be seen to be stable without control, reaching a steady state at a head determined by the physical dimensions of the system expressed as capacitance and resistance. The only requirement for stability is that the single root be negative, which it is. The system is said to be 'inherently' stable and this implies that those energy conversions that take place within the system are non-reversible. Energy exists in this system in three forms, potential or head energy, kinetic or velocity energy, and frictional energy (of the turbulence created by the resistance of the valve). Since the resistance R is constant the velocity through the valve is a dependent variable of head

$$\frac{\phi}{h} = R \quad \text{but } \phi = v \times A \quad \text{hence } \frac{v}{h} = \frac{R}{A} \tag{3.9}$$

where A is the cross-sectional area through the valve and is also constant.

Hence $\phi = \text{constant} \times h$.

From a dynamic viewpoint head energy and velocity energy are totally dependent and the only energy conversion taking place between independent forms is from velocity to frictional energy which is irreversible.

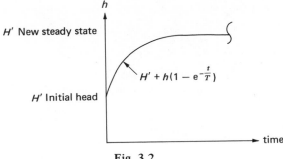

Fig. 3.2.

The change from any stable state to another, when a step change of input is applied, will be of exactly the same form, the initial values of I and ϕ being other than zero (Fig. 3.2).

3.2 Transfer functions including control

In a real industrial system it is normally required that the head in a vessel be controlled at some predetermined level regardless of variation of the inflow rate I. To this end the error, or difference between the head and the desired head, is measured, amplified, and fed back negatively. The block diagram for this system is as shown in Fig. 3.3.

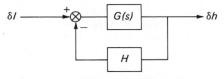

where H is the amplifier gain.
Fig. 3.3.

The transfer function

$$\frac{\delta h(s)}{\delta I(s)} = \frac{G(s)}{1 + G(s).H} = \frac{\dfrac{1/C}{(s + 1/CR)}}{1 + \dfrac{1/C}{(s + 1/CR)} H} = \frac{1/C}{s + \left(\dfrac{1 + RH}{CR}\right)} \quad (3.10)$$

δh and δI are used to show that these are the changes in the variables not the absolute values. The object of the system then is to reduce δh to a minimum for any δI. From the transfer function above it can be seen that the characteristic equation is

$$s + \frac{1+RH}{CR} = 0$$

and since the input forcing function is δI the differential equation is

$$s(\delta h) + \frac{(1+RH)\delta h}{CR} = \frac{\delta I}{C}$$

the steady-state solution: $(1+RH)\delta h = R\delta I$.

Now $\delta h = (h - h_d)$, actual head minus desired head, so in the new steady state

$$(1+RH)(h - h_d) = R\delta I$$

or

$$\frac{R}{(1+RH)} \delta I = (h - h_d) \qquad (3.11)$$

From this result it can be seen that if H is zero (no feedback), $(h - h_d) = R\delta I$ as previously, but if H is large, the term on the left becomes small and

$$h = h_d \text{ in the limit as } H \to \infty$$

so that the error δh is reduced by increasing the gain H.

The transient solution which, without control, was

$$h = k\, e^{-t/T}$$

becomes with control

$$h = k\, e^{-(1+RH)/RC\, t} \qquad (3.12)$$

and it is obvious that it will decay more rapidly as H increases. It can be seen that the addition of a simple proportional (fixed gain H) control reduces the error between the required constant head and the actual head under conditions of changing inflow rate by varying the resistance at the outlet; the term $(1+RH)/RC$ being made up of $1/RC$ and H/C of which the first is the original time constant term and the second an addition (positive or negative) to it. This is achieved by adding a motor and positioning mechanism to the throttling value as shown in Fig. 3.4. Since the higher the gain H the smaller the error *and* the faster the approach to the new steady state after a change of I (faster decay of transient response), it is obvious that in theory the highest gain will give the best results. Two things limit this in a practical case. First, the gain H depends not only on the amplifier in the controller, but on the physical dimensions of the controlling valve and lines; it being only possible to increase or decrease the resistance within certain limits.

The second reason that, in practice, the gain must be limited is that ideal control mechanisms do not exist. Elements of the control mechanism may

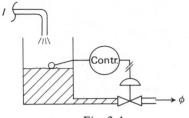

Fig. 3.4.

introduce other time constant elements into the system thus making possible oscillatory and even unstable response. For instance, the motor of the control valve will probably comprise a diaphragm chamber, the valve being positioned by compressed air. A comparison of the transfer functions of the system without control

$$\frac{1/C}{\left(s + \dfrac{1}{CR}\right)}$$

and the system with control

$$\frac{1/C}{\left(s + \dfrac{1+RH}{CR}\right)} \quad \text{or} \quad \frac{1/C}{\left(s + \dfrac{1}{CR/1+RH}\right)}$$

shows that the addition of control has reduced the process time constant from CR to $CR/(1+RH)$. The time constant of the control valve must be small in comparison to that of the process for the system to behave in a first-order fashion. As H is increased and the system time constant decreases, the relative significance of the control valve transfer function is increased and eventually the system will behave as a second- or (if there are other time constants) higher-order system, and may become unstable to fast changing inputs as shown in

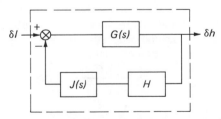

$J(s)$ is the transfer function of the control mechanism.

$$J(s) = \frac{1/C'}{\left(s + \dfrac{1}{T'}\right)}$$

Fig. 3.5.

Fig. 3.5. The system transfer function

$$\frac{\delta h(s)}{\delta I(s)} = \frac{G(s)}{1 + G(s) \cdot J(s) \cdot H}$$

$$= \frac{\frac{1/C}{(s + 1/T)}}{1 + \frac{1/C \cdot 1/C_1}{\left(s + \frac{1}{T}\right)\left(s + \frac{1}{T'}\right)} \cdot H}$$

where T' is the time constant of the control mechanism and C' the capacitance of the control mechanism,

$$= \frac{1/C \left(s + \frac{1}{T'}\right)}{\left(s + \frac{1}{T}\right)\left(s + \frac{1}{T'}\right) + \frac{H}{CC_1}}$$

and

$$s^2 + \left(\frac{1}{T} + \frac{1}{T'}\right)s + \left(\frac{1}{TT'} + \frac{H}{CC'}\right) = 0 \qquad (3.13)$$

is seen to be the characteristic equation.

In factored form

$$(s + a)(s + b) = 0$$

where

$$(a + b) = \left(\frac{1}{T} + \frac{1}{T'}\right) = 2\zeta\omega = k_1$$

and

$$ab = \left(\frac{1}{TT'} + \frac{H}{CC'}\right) = \omega^2 = k_2$$

If $4k_2 > k_1^2$, the roots become complex and due to the presence of H in the term k_2 this is obviously possible. The stability of the system will depend on the value of the time constant/s introduced by the control valve, and the gain H. Contrast this with the second-order system comprising two tanks as shown in Fig. 3.6. The transfer function of each is

$$\frac{1/C}{\left(s + \frac{1}{T}\right)}$$

and since they are in series the overall T.F. is

$$\frac{\delta h(s)}{\delta I(s)} = \frac{1/CC_1}{\left(s + \frac{1}{T_1}\right)\left(s + \frac{1}{T_2}\right)} \qquad (3.14)$$

of which the roots $s = -1/T_1$ and $s = -1/T_2$ are wholly real. They must be real since the energy conversions which take place in this system are all irreversible. The addition of feedback control then makes possible oscillatory and, in extreme cases, unstable response. It must therefore have introduced reversible energy conversion where none existed before. The only physical change made to the system when control was added was to motorize the valve, thus making its resistance variable. As a result, since $v = (R/A) \times h$ and (R/A) is no longer constant, velocity energy and head energy are no longer independent, but can be interchanged by the variation of (R/A). Velocity energy could not be converted to head energy in the system without feedback because gravity cannot be defied and water therefore cannot flow uphill. By varying (R/A) this conversion is made possible. Since oscillation implies complex roots to the characteristic equation and complex roots imply a phase or time lag in the energy feedback, it is also necessary to have a second time constant element in the feedback path for oscillatory response to be possible.

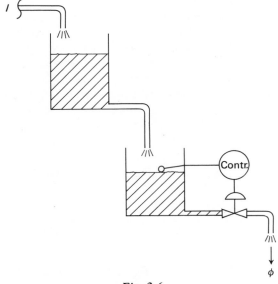

Fig. 3.6.

In practice, it is the control valve motor capacity, together with the resistance of compressed air lines and fittings, which introduce a second time constant in the feedback path, making instability possible. This time constant is fairly easily reduced by adding a booster relay which effectively reduces R, and thus T, so that high gain may be used without oscillation, and the error δh is kept small. Simple proportional control (H = gain) is therefore quite suitable for the control of level in most cases.

Unfortunately the same cannot be said for most thermal processes where the measurement often introduces one or more quite large RC time constants. The cause of instability is almost invariably the measurement, not the control valve, and is not easily corrected. The only remedy is to reduce the heat transfer time constants as far as possible, but even so it is often not possible to use high gain and a large error occurs. The basis on which the control mechanism can be altered to compensate for this and reduce the error will be discussed in chapter 5.

3.3 Inertia in the process

Systems comprising capacitative/resistive elements in which compressible fluid flow takes place, such as gas compressor pipeline and vessel systems, constitute a process which, without external feedback, can oscillate, though not become unstable in the strict sense of the word. Energy exists in such a system as pressure energy, kinetic energy, and friction. Unlike the earlier example of liquid level, the conversion from pressure energy to velocity energy is reversible within the system as head to kinetic energy in the pendulum. Energy feedback can take place within the system and the inertia of the fluid introduces a time constant which must be taken into account whenever velocity and head or pressure energy forms are independently variable. This inertia effect is present in the level system, but can usually be neglected as it is 'swamped' by the resistive capacitative elements. Inertia is analogous to inductance in an electrical circuit.

Since the inertia time constant is in the energy feedback path, it introduces phase lag as the RC time constant in the external feedback path of the level control does.

Defining L the inertia, as the head required to support unit rate of change of outflow,

$$\text{the dimensions of } L \text{ are } \frac{\text{length}}{\text{length cubed/time squared}}$$

$$\text{and those of } R \text{ are } \frac{\text{length}}{\text{length cubed/time}}$$

Therefore the dimension of L/R is time, and this is the new time constant. The basic equation

$$sh = \frac{1}{C}(I - \phi)$$

for fluid flow into and out of a capacitance or vessel becomes

$$sh = \frac{1}{C}\left(I - \frac{h}{R} - \frac{h}{LS}\right) \qquad (3.15)$$

since
$$\phi = \frac{h}{R} + \frac{h}{LS}$$

hence
$$hs^2 + \frac{h}{RC}s + \frac{h}{LC} = \frac{I}{C}s$$

and
$$\frac{h(s)}{I(s)} = \frac{\frac{1}{C}s}{\left(s^2 + \frac{1}{CR}s + \frac{1}{LC}\right)} \qquad (3.16)$$

The characteristic equation is
$$s^2 + \frac{1}{CR}s + \frac{1}{LC} = 0$$

which can be re-written
$$s^2 + \frac{1}{CR}s + \left(\frac{R}{L}\frac{1}{CR}\right) = 0$$

[P.43 T = time const. = CR]

or
$$s^2 + \frac{1}{T}s + \frac{1}{T}\cdot\frac{1}{T'} = 0 \qquad (3.17)$$

where $T' = L/R$ the inductive TC.

It can be seen that the roots of the characteristic equation are
$$-\frac{1}{2T} \pm \sqrt{\left(\frac{1}{T^2} - \frac{4}{TT'}\right)}$$

If $L = \frac{1}{4}R$ this becomes
$$-\frac{1}{2T} \pm \sqrt{\left(\frac{1}{T^2} - \frac{1}{T}\right)}$$

and if CR is greater than unity $1/T > 1/T^2$, and the roots are complex, leading to oscillatory response. If CR is less than unity and $L = \frac{1}{4}R$, the roots will be wholly real and no oscillation will occur. If these relationships are maintained, $CR < 1$ and C is very large, while R is very small, L will be very small also (to give $L = \frac{1}{4}R$), which is to say that there is little inertia and little damping R. In these circumstances the oscillations will be very large, but notice that actual instability cannot occur as $1/2T$ is always greater than zero.

An example of the above is to be found in the phenomenon of compressor surge. Here the interchange of velocity energy with pressure energy between the rotating impellor and the stationary diffuser takes place with virtually no resistance, little inertia, and large capacitance. Due to the increasing mismatch of blade angles and diffuser vanes as the throughput or outflow decreases the efficiency of velocity to pressure conversion decreases and, as shown in Fig. 3.7, the pressure becomes increasingly 'de-coupled' from the flow. It is as

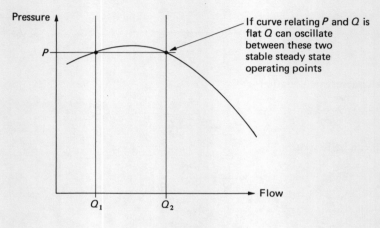

Fig. 3.7.

though the level in the tank in the earlier example were not affected at all by changes in the outflow. The system then is in a highly oscillatory condition and has virtually no self-regulation at all. The momentum changes, as flow oscillates violently, impose oscillatory or vibratory stresses on the bearings and impellors, and in an axial machine surge these stresses are almost always disastrous.

It has been demonstrated that the order of the characteristic equation which determines the transient behaviour of any system is equal to the number of effective time constants. Whether the system oscillates depends on the reversibility or otherwise of the energy forms associated with these time constants. The two time constants of the double tank level example lead to 'irreversible' behaviour because they are both associated with the same form of energy and are in series, the output of one being the input of the other.

The single tank with control also contains two time constants, but in this case they are in parallel and 180° out of phase, thus permitting reversible energy exchange: any reduction of outflow will cause an increase of head and any increase of outflow a decrease—hence flow and head are 180° out of phase.

Returning to consideration of compressible fluid flow this time with simple 'proportional' feedback control added

$$\frac{h(s)}{I(s)} = \frac{G(s)}{1+G(s)H} = \frac{\dfrac{1}{\dfrac{C}{s}\left(s^2 + \dfrac{1}{T}s + \dfrac{1}{LC}\right)}}{1 + \dfrac{H}{\dfrac{C}{s}\left(s^2 + \dfrac{1}{T}s + \dfrac{1}{LC}\right)}}$$

$$= \frac{s}{C\left[s^2 + \left(\dfrac{1}{T}+H\right)s + \dfrac{1}{LC}\right]} \qquad (3.18)$$

for which the characteristic equation is

$$s^2 + \left(\frac{1}{T}+H\right)s + \frac{1}{LC} = 0 \qquad (3.19)$$

It can be seen that H appears not in the last term but in the first-order term, so that damping is increased by increase of gain, not decreased.

The above transient response is based on the assumption that control is instantaneous, but the true state of affairs is normally complicated by the existence of a significant RC time constant in the feedback control path attributable to the control valve motor. Hence

$$\frac{h(s)}{I(s)} = \frac{G(s)}{1+G(s)H(s)} = \frac{\dfrac{s}{C\left(s^2 + \dfrac{s}{T} + \dfrac{1}{LC}\right)}}{1 + \dfrac{K}{C\left(s^2 + \dfrac{s}{T} + \dfrac{1}{LC}\right)\cdot\left(s + \dfrac{1}{T}\right)}}$$

where T' is an RC time constant of the control mechanism and

$$H(s) = \frac{K}{\left(s + \dfrac{1}{T'}\right)} \qquad (K = \text{gain})$$

Hence

$$\frac{h(s)}{I(s)} = \frac{s\left(s + \dfrac{1}{T'}\right)}{\left(C\,s^2 + \dfrac{s}{T} + \dfrac{1}{LC}\right)\left(s + \dfrac{1}{T'} + K\right)}$$

or

$$\frac{h(s)}{I(s)} = \frac{s\left(s + \frac{1}{T'}\right)}{C\left[s^3 + \left(\frac{1}{T} + \frac{1}{T'}\right)s^2 + \left(\frac{1}{LC} + \frac{1}{TT'}\right)s + \left(\frac{K}{C} + \frac{1}{T'LC}\right)\right]}$$

and the characteristic equation of the closed-loop system is

$$s^3 + \left(\frac{1}{T} + \frac{1}{T'}\right)s^2 + \left(\frac{1}{LC} + \frac{1}{TT'}\right)s + \left(\frac{K}{C} + \frac{1}{T'LC}\right) = 0$$

or*

$$s^3 + \left(\frac{1}{T} + \frac{1}{T'}\right)s^2 + \left(\frac{1}{TT_L} + \frac{1}{TT'}\right)s + \left(\frac{K}{C} + \frac{1}{TT_LT'}\right) = 0 \quad (3.20)$$

where T_L is the process inertia time constant.

Clearly the addition of a control mechanism has increased the complexity of the closed-loop system.

The process transfer function for incompressible flow is

$$G(s) = \frac{h(s)}{I(s)} = \frac{s(1/C)}{\left(s^2 + \frac{s}{CR} + \frac{1}{LC}\right)}$$

and the characteristic equation is therefore

$$s^2 + \left(\frac{s}{CR}\right) + \left(\frac{1}{LC}\right) = 0$$

However C is almost zero and so comparing this with the form

$$s^2 + 2\zeta\omega s + \omega^2 = 0$$

it can be seen that the frequency of any oscillation will be very high, while the damping factor ζ (and hence any tendency to oscillate) will depend on the relative values of R and L and is normally very large. Adding feedback control, as for the compressible flow case, the characteristic equation

$$Cs^3 + \left(\frac{1}{R} + \frac{C}{T'}\right)s^2 + \left(\frac{1}{L} + \frac{1}{R.T'}\right)s + \left(K + \frac{1}{T'L}\right) = 0$$

tends to

$$\left(\frac{1}{R}\right)s^2 + \left(\frac{1}{L} + \frac{1}{RT'}\right)s + \left(\frac{1}{T'L} + K\right) = 0 \quad (3.21)$$

as $C \to$ zero.

---

* $\frac{1}{LC} = \frac{R}{L} \cdot \frac{1}{CR} = \frac{1}{T_L T}$

This can be re-written

$$s^2 + \left(\frac{1}{T_L} + \frac{1}{T'}\right)s + \left(\frac{1}{T'T_L} + RK\right) = 0$$

The roots of this equation are

$$s = -\frac{1}{2}\left(\frac{1}{T_L} + \frac{1}{T'}\right) \pm \frac{1}{2}\sqrt{\left\{\left(\frac{1}{T_L} + \frac{1}{T'}\right)^2 - 4\left(\frac{1}{T_L T'} + RK\right)\right\}} \quad (3.22)$$

and if $T_L = T' = \textcircled{T}$.

$$s = -\frac{1}{\textcircled{T}} \pm \frac{1}{2}\sqrt{(RK)}$$

Thus, if the control time constant equals the process inertia time constant closed-loop oscillation increases with K the controller gain. If there is no feedback, $K = 0$ and the roots are real. When these two time constants are of very different size, the first term under the root sign in (3.22) increases relative to the second and non-oscillatory response is possible with small values of the gain K.

3.4 Inertia in the control mechanism

The inertia time constant of the control mechanism cannot always be ignored as it has been until now. Let

$$G(s) = \frac{s}{C\left(s^2 + \dfrac{s}{RC} + \dfrac{1}{LC}\right)}$$

as previously, and

$$H(s) = \frac{Ks}{C'\left(s^2 + \dfrac{s}{R'C'} + \dfrac{1}{L'C'}\right)} \quad (3.23)$$

where R', C', and L' refer to the control mechanism. Hence

$$\frac{G(s)}{1 + G(s)H(s)} = \frac{s\left(s^2 + \dfrac{1}{R'C'}s + \dfrac{1}{L'C'}\right)}{C\left(s^2 + \dfrac{s}{RC} + \dfrac{1}{LC}\right)\left(s^2 + \dfrac{s}{R'C'} + \dfrac{1}{L'C'}\right) + \dfrac{Ks^2}{C'}} \quad (3.24)$$

and the characteristic equation is

$$C\left[s^4 + \left(\frac{1}{RC} + \frac{1}{R'C'}\right)s^3 + \left(\frac{1}{LC} + \frac{1}{L'C'} + \frac{1}{RC.R'C'} + \frac{K}{C'C}\right)s^2 \right.$$
$$\left. + \left(\frac{1}{LC.R'C'} + \frac{1}{L'C'.RC}\right)s + \frac{1}{LC.L'C'}\right] = 0$$

or

$$Cs^4 + \left(\frac{1}{R} + \frac{C}{R'C'}\right)s^3 + \left(\frac{1}{L} + \frac{C}{L'C'} + \frac{1}{R}\cdot\frac{1}{R'C'} + \frac{K}{C'}\right)s^2$$
$$+ \left(\frac{1}{L}\cdot\frac{1}{R'C'} + \frac{1}{L'C'}\cdot\frac{1}{R}\right)s + \frac{1}{L}\cdot\frac{1}{L'C'} = 0$$

Let C tend to zero then the characteristic equation tends to

$$s^3 + \left(\frac{R}{L} + \frac{1}{R'C'} + \frac{RK}{C'}\right)s^2 + \left(\frac{R}{L}\cdot\frac{1}{R'C'} + \frac{1}{L'C'}\right)s + \left(\frac{R}{L}\cdot\frac{1}{L'C'}\right) = 0$$

Which can be re-written

$$s^3 + \left(\frac{1}{T_L} + \frac{1}{T'} + \frac{RK}{C'}\right)s^2 + \left(\frac{1}{T_L T'} + \frac{1}{T'T'_L}\right)s + \left(\frac{1}{T_L T'T'_L}\right) = 0 \quad (3.25)$$

where T_L is the process inertia time constant, T' is the control mechanism RC time constant, and T'_L is the inertial time constant of the control. If T_L becomes *relatively* very large tending towards ∞, the equation tends towards

$$s^2 + \left(\frac{1}{T'} + \frac{RK}{C'}\right)s + \left(\frac{1}{T'T'_L}\right) = 0 \quad (3.26)$$

the roots of which are

$$s = -\frac{1}{2}\left(\frac{1}{T'} + \frac{RK}{C'}\right) \pm \frac{1}{2}\sqrt{\left\{\left(\frac{1}{T'} + \frac{RK}{C'}\right)^2 - \frac{4}{T'T'_L}\right\}} \quad (3.27)$$

If T'_L is small the term under the root sign may be negative and response oscillatory. Unfortunately the addition of a position servo to overcome gland friction has the effect of reducing T'_L since the servo, normally operating at high gain, increases the valve positioning speed thus reducing inertia in the control mechanism. In a normal liquid flow system the inertia L is large and the resistance R fairly small giving a very large process inertia time constant.

Interchange of energy takes place between pressure in the diaphragm of the valve motor and momentum of process fluid by virtue of 180° phase difference between the two forms, exactly as in the compressible flow case where oscillation is known as 'surge'. The control mechanism turns the system into an oscillator: this does not matter if the process RC time constants are large as in the case of level control, as the process provides ample damping, but in the case of liquid flow control it does not. The addition of a booster relay reduces R' and, hence T', thus increasing the value of the positive term under the square root sign relative to the negative term; this reduces the tendency to oscillation.

3.5 Conclusion

The examples covered in this chapter are all taken from industrial process control practice, partly because of the author's background and partly because the illustrations tend to have common ground with most engineer's experience. The reader should understand, however, that the principles involved are universal. A system can be defined as a group of component parts all of which interact. In its simplest form (for instance, a system of 'weightless' levers) transient response is instantaneous and the only relationships of interest are the steady state ones. In most engineering systems, however, there are energy state changes within the system, and transient behaviour, especially where oscillation is possible, is of very great importance. Except for a few elements, such as fluid flow and some chemical reactions, it is often the addition of feedback control which introduces oscillation. The majority of industrial process elements are inherently stable, and actual instability is comparatively rare except in the case of speed control of moving machinery where inertia time constants tend to be very significant, leading to high-order characteristic equations. The skill of the engineer lies in identifying the elements which are significant and those which are not, for it is rare to be able to quantify all the information required to solve the characteristic equations.

Much analysis and sound design can be carried out, from whatever information is available, by an engineer who understands the principles of dynamic response sufficiently well to be able to apply them to his own technology. Consider, for instance, the economy of the country: money and labour are energy forms existing within the system; their exchange is reversible. The inertia in the system is caused by the capital equipment and stability of labour requirements for production, since there are forces acting against sudden change of rate of production. Damping is largely applied by the manipulation of taxes which are the resistance element in the system. In more recent times, restriction of the available money supply, in other words change of capacity, has been used to regulate damping. As has been seen in this chapter this does not necessarily have the same effect as changing the taxes or resistance. The problem in applying the systems approach to economics lies in the difficulty of making accurate and fast measurements. This will perhaps be solved by a better understanding of statistical mathematics, but until it is the timid 'touch on the regulator' not the full-blooded application of systems science will be the order of the day.

4. Basic principles of design of systems

4.1 Introduction to design

The design process involves definition, selection, and synthesis in that order, though often the process must be iterative. The first task in any design is to define the elements with sufficient degree of resolution to serve the purpose in hand. The second task is to define the necessary relationships between elements. As has been seen already an element of the particular system under consideration may itself be a large and complex system; for example, one element of a chemical plant might be a refrigeration plant to supply cooling water for various purposes within the chemical plant system. The designer of the chemical plant will regard this as an element of his system, while to the designer of the refrigeration plant it is the system. The chemical plant designer is interested in defining the relationship between what goes into and what comes out of the refrigeration plant—in systems terminology the 'relating function' of 'responses' to 'stimuli'. For his purposes, it is sufficient to enclose the refrigeration system within a boundary or 'interface' (Fig. 4.1).

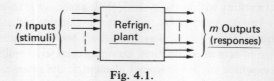

Fig. 4.1.

It is then his responsibility to define the n stimuli or inputs and m responses he requires in sufficient detail to enable the refrigeration unit designer to define his system elements and relationships so that it has a relating function which is adequate. He must remember that the stimuli include *all* effects and influences of the environment on the refrigeration plant and that the environment includes the rest of the chemical plant *and other outside elements.*

At the beginning of the design of any system the definition of some elements will already have been established. This may be so because existing plant must be used, or it may be that some major element can only be made to a certain design, or for a variety of other reasons. In designing a system, we are concerned with the *behaviour* of the elements which comprise it, and there are basically only two ways in which the behaviour pattern or 'relating function' can be established. The first method is calculation, the second test. In order to establish the RF (relating function) of the elements which are to be designed, the first method is essential, but at the same time some approximations and simplifications are usually possible. The RF of an existing element, however, may not be calculated easily and often test methods are used. Sometimes the RF of a major element of process plant is itself time dependent, as, for instance, when catalysts are used to promote chemical reactions; in such cases it may be necessary to measure and define the RF continuously during operation of the plant and modify the relationships between this and other elements accordingly.

Just how difficult it can be to establish a relating function for a system can be appreciated by considering the very simple example, Fig. 4.2.

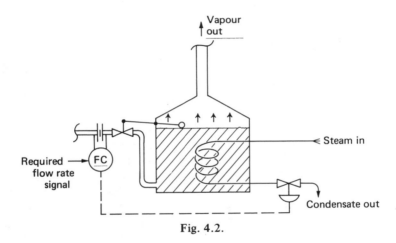

Fig. 4.2.

A boiler, using steam, heats a low vapour point liquid to evaporate a given quantity per unit time. In order to do so, the flow of heat energy into the system must be maintained in constant ratio to the flowrate of vapour. The boiling liquid is at one constant temperature and the steam at another. Hence to supply a given quantity of heat into the system a corresponding heat transfer area will be required; it is the heat transfer area which must change to effect control. The system operates by restricting the entry of steam through the valve so that just enough enters to supply the required heat; the pressure in the coil is slightly below atmospheric. Thus, pressure, plus the head of condensate in the coil, equals atmospheric pressure. The area of the coil available for heat transfer from condensing steam is determined by the head of condensate and therefore indirectly by the throttling effect of the control valve. (Assuming that the coil is not very high, the slight difference in condensing pressure with varying heads of condensate make very little difference to the condensing temperature. Therefore, provided the boiling temperature of the process liquid is much lower than this temperature, it is a reasonable approximation to say that the condensing temperature is constant.)

Now suppose that the required rate of vapour flow increases, necessitating a greater heat transfer area, i.e., a lower head of condensate in the coil. The outlet from the coil is unlikely to be restricted and a very rapid change of condensate head is possible. If, however, a decrease in vapour rate is called for, the rate at which the heat transfer area changes depends on the rate of condensation of the steam at that time and the physical dimensions (surface area/volume) of the coil tube, as head can only increase as condensate accumulates.

It can already be seen that the relating function of *stimulus* (change of vapour flowrate requirement) and *response* (change of vapour flowrate) is far from simple and is, in fact, asymmetrical in the sense that the response to an increase in stimulus is quite different to the response to a decrease.

4.2 Analysis by 'frequency response'

In chapter 2 it was pointed out that the response of any system to sinusoidal 'forcing' can only be another sinusoid since purely imaginary (oscillatory) functions cannot give rise to 'real' (steady-state) response. As a result the particular solution of the describing differential equation is obtained, not by setting S, the complex operator, equal to zero, but by making $s = j\omega$.

The Fourier transform of the response function is a complex combination of sinusoidal 'elements' so that the response to any particular sinusoid of frequency ω is a complex quantity which relates input to output in terms of phase angle and magnitude, the *form* of both being necessarily identical. The response can be expressed as a function of $j\omega$ therefore, or (since j is invariable) of ω the forcing frequency. Thus, the most obvious way of obtaining the transfer function or transient response of a system by test is to force it with sinusoidal inputs of a *suitable* range of frequencies, recording phase lag and attenuation at each separate test for each frequency. The result, a function of $j\omega$ can be translated into the general transient response by replacing the purely imaginary (oscillatory) operator $j\omega$ by the general complex operator s to give the Laplace integral transform.

In the course of the first chapter a system was defined as a collection of component parts each of which interacts with others. It has been demonstrated that the dynamics of the interaction can be described algebraically by use of the operator notation and the concept of integral transforms. If two elements are in series as shown in Fig. 4.3(a), the input/output relationship is $G_1(s) \times G_2(s)$, and, if two elements are in parallel as in Fig. 4.3(b), the relationship is

$$\frac{G(s)}{1 + G(s)H(s)}$$

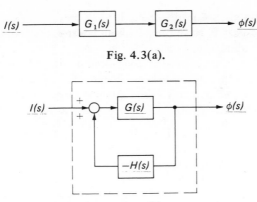

Fig. 4.3(a).

Fig. 4.3(b).

which, if $G(s)$ and $H(s)$ are both *first*-order terms, gives a *second*-order transfer function and can be treated as a block. Thus $G_2(s)$ in Fig. 4.3(a) might comprise 4.3(b): the total transfer function $G_1(s) \cdot G_2(s)$ would then be third-order.

In a real system any third-order transfer function must contain one real root, i.e., one first-order element, since complex roots can only occur in conjugate *pairs* or not at all (a physical system, unlike an input or output function *cannot* comprise an imaginary component). A fourth-order, or any even order system, can have all complex roots and can be represented by multiples of second-order elements. Thus, it can be deduced that *any* real system response function can be broken down into a strictly limited number of elements

1. A constant ratio K as for a weightless lever system.
2. A derivative s or integral $1/s$ component.
3. A first order $(s + 1/T)$ component.
4. A second order $(s^2 + 2\zeta\omega s + \omega^2)$ component.

Moreover, any system transfer function will comprise *multiple combinations* of such elements so that, if $\log_e$ of the attenuation is plotted against ω, the elements will be combined additively, which is to say that they are superimposed on each other in graphical terms. From a plot of attenuation against frequency obtained from tests on a system, it is often possible to synthesize an approximation to the transfer function by identifying the four elements above in the function obtained. The phase/frequency plot can be superimposed on the attenuation plot or can be separate and any identification of the four basic elements must, of course, agree with the phase/frequency plot.

The synthesized transfer function obtained by these means can be used for analysis of system behaviour, but the basic assumption that the system is essentially linear and time invariant must never be forgotten.

The transfer function of most industrial process plants are almost impossible to calculate, but the transfer function of the control loop can usually be estimated fairly accurately and in any case is often 'tuneable' on site. Hence, if the plant is tested and its T.F. established without control, the control requirements can be readily established.

Since phase angle is not expressed logarithmically it is useful if attenuation is converted to log form and plotted on a linear scale so that both plots can be made together if desired. For this reason attenuation or magnitude ratio is usually expressed in decibels

$$m = 20 \log_{10} M$$

Mag ratio M 1:1 $m = 20 \times (0) = 0$ dB.
M 0.1:1 $m = 20 \times (\bar{1}.0) = 20 \times (-1) = -20$ dB.
M 0.2:1 $m = 20 \times (\bar{1}.30D) = 20 \times (-0.7) = -14$ dB.
M 0.4:1 $m = 20 \times (\bar{1}.600) = 20 \times (-0.4) = -8$ dB.

Notice that a doubling or halving of the attenuation is equivalent to an increase or decrease of 6 dB.

The plots for attenuation and phase response of each of the four basic elements are shown in Fig. 4.4. Considering the first-order element $1/(s + 1/T)$ the response to sinusoidal forcing is given by replacing s with $j\omega$, thus

$$\frac{\phi(j\omega)}{I(j\omega)} = \frac{1}{j\omega + 1/T} = \frac{T}{(jT\omega + 1)} \qquad (4.1)$$

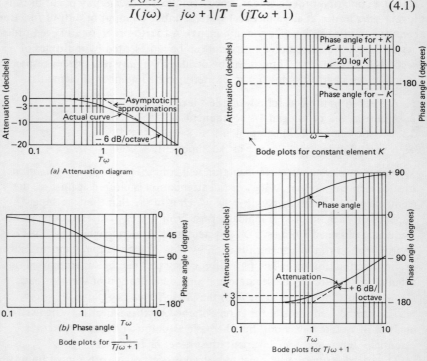

Fig. 4.4.

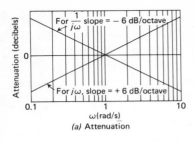
(a) Attenuation

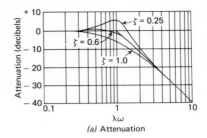
(a) Attenuation

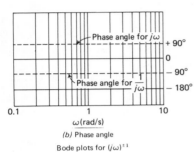
(b) Phase angle

Bode plots for $(j\omega)^{\pm 1}$

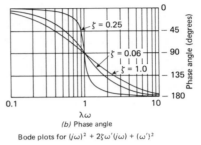
(b) Phase angle

Bode plots for $(j\omega)^2 + 2\zeta\omega'(j\omega) + (\omega')^2$

Fig. 4.4.—*continued.*

$(jT\omega + 1)$ is a complex function of ω—that is there is a complex value of $(jT\omega + 1)$ for any value of ω. Re-writing the complex value in alternative form

$$r \angle \theta \tag{4.2}$$

where

$$r = [(T\omega)^2 + 1]^{1/2} \tag{4.3}$$

and

$$\theta = \tan^{-1}\frac{T\omega}{1} = \tan^{-1}T\omega \tag{4.4}$$

hence

$$\frac{\phi(j\omega)}{I(j\omega)} = \frac{1}{r \angle \tan^{-1}(T\omega)} = \frac{1}{r} \cdot \angle - \tan^{-1}(T\omega) \tag{4.5}$$

where $\frac{1}{r}$ is the attenuation and $-\tan^{-1}(T\omega)$ is the phase angle.

If

$$T\omega = 0, \quad \frac{1}{r} = \frac{1}{1} = 1 \quad \text{and} \quad \angle -\tan^{-1}(0) = 0°$$

and if

$$T\omega = 1, \quad \frac{1}{r} = \frac{1}{\sqrt{2}} = 0.71 \quad \text{and} \quad \angle -\tan^{-1}(1) = -45°$$

Other values of the attenuation $1/r$ and the phase $\angle\theta$ can be calculated to give the diagram in Fig. 4.4. Notice that the actual curve is asymptotic to the zero attenuation line at one extreme and a line of fixed slope -6 dB/octave or -6 dB per doubling of the frequency at the other. This is because, as ω becomes very small $[(T\omega)^2 + 1]^{1/2}$ tends towards unity giving no attenuation, (0 dB) and as ω becomes very large $[(T\omega)^2 + 1]^{1/2}$ tends towards $T\omega$ so that the attenuation $1/r$ is $1/T\omega$ and doubles (-6 dB) for every doubling of the frequency. Similarly, the element s in the transfer function becomes $T\omega$ and $1/s$ becomes $1/T\omega$ so that these are represented by straight lines having a slope of ± 6 dB/octave.

It will be seen that a family of curves, not just one curve, is obtained for any second-order element, each curve representing a different value of ζ the damping ratio. For values of $\zeta \geqslant 1.0$ the curve obtained is similar to the $(Ts + 1)$ first-order element with the one difference that the slope of the line to which it is asympototic, as ω becomes larger, is not ± 6 dB but ± 12 dB. This represents a second-order element with wholly real roots (no oscillation), and as $\omega \rightarrow \infty$

$$(jT\omega + 1).(jT\omega + 1) \rightarrow (T\omega)^2$$

In other words, it tends towards a double differentiation (s^2) term hence the ± 12 dB slope.

As ζ decreases and the response becomes oscillatory the form of the curve alters, giving at or close to the $\omega = \omega'$ point, not attenuation, but amplification and as $\zeta \rightarrow 0$ this amplification tends to ∞. We have come across this phenomenon already (eqn. (2.24)) in the case of compressible flow and in particular compressor surge, and it is well known as 'resonance' to radio engineers who deliberately design 'tuned' circuits in order to utilize the amplification which results. The natural frequency ω' of the unforced element is a factor which does not occur in the other basic units since they are not capable of oscillation on their own.

It will be noticed that the asymptotes themselves intersect the zero dB line at values $\omega = 1/T$ for a first-order element and $1/\beta$ or ω' for a second-order element. This gives a very useful guide to plot the 'Bode diagrams' as these logarithmic plots are called.

In addition to identifying potential oscillation of a system, the Bode diagram can be used to determine whether or not the system can behave in an unstable fashion. As has been seen in earlier chapters oscillatory behaviour of a system is the result of bi-directional or reversible energy interchange within the system. When the two energy forms concerned are $180°$ or (-1) out of phase an increase in one is accompanied by a corresponding and equal decrease in the other. Such is the case with a compressor in surge; pressure and momentum energy being exactly $180°$ out of phase. The resultant oscillations do not increase the steady-state or mean energy output at all, and the system is never unstable in the strict mathematical meaning of the word; damage results from oscillation not instability! In the general case, however, energy can be 'fed back' at phase angles other than 180°, one energy form increasing at a greater rate than the other decreases and the total energy within the system therefore increases with time. Unless such net increase is offset by a non-reversible energy loss within the system, such as friction, the system will be unstable and one of the roots of the describing characteristic equation will have positive real parts.

Negative feedback control implies the introduction of energy feedback 180° out of phase to change steady-state relationships. As has been seen, however, the existence of time constants and pure time delays within the system has the effect of increasing the phase lag. At a certain frequency the output will be, not 180° (−1) out of phase with the input but $360° \equiv 0°$ (+1). In other words the output will be in phase with the input and an increase in one energy form within the system will be accompanied, not by a corresponding decrease in another, but by a corresponding increase. Attenuation or damping due to non-reversible energy changes such as friction may be sufficient, even under these conditions, to ensure a nett decline in energy in the system and render it stable in which case the system will always be stable. The criteria of stability then, is that the gain must be less than 0 dB when the output is lagging the input by 180° (and the energy feedback is therefore 180° + 180° = 360° out of phase).

It can safely be assumed that higher forcing frequencies will give greater attenuation and thus reinforce the input less. Figure 4.5 shows two attenu-

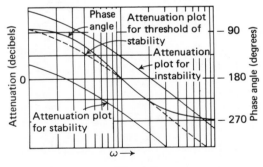

Fig. 4.5.

ation plots and one phase plot. A system described by the upper attenuation plot and the phase plot would be unstable, but one described by the lower attenuation plot together with the same phase plot would be stable.

4.3 Analysis by phase/gain plots

The attenuation and phase 'Bode' plots are not the only way in which the results of a series of frequency forcing tests of a system can be plotted. Since each test yields a complex number (magnitude ratio and phase angle) the locus of such values as the forcing frequency varies from zero to infinity can be plotted on polar coordinates as shown in Fig. 4.6.

FREQUENCY RESPONSE DATA

ω(radians/sec)	$M(\omega)$	$\theta(\omega)$(degrees)
0	1.00	0.0
2	0.98	−11.3
5	0.89	−26.6
10	0.71	−45.0
20	0.45	−63.4
40	0.24	−76.0
∞	0.00	−90.0

NYQUIST DIAGRAM FOR FIRST ORDER COMPONENT

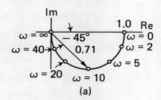

(a)

FREQUENCY RESPONSE DATA

ω(radians/sec)	$M(\omega)$	$\theta(\omega)$(degrees)
0	1.00	0.0
2	1.02	−11.8
5	1.11	−33.7
8	1.14	−65.8
10	1.00	−90.0
12	0.78	−110.1
15	0.51	−129.8
20	0.28	−146.3
40	0.06	−165.1
70	0.02	−171.7
∞	0.00	−180.0

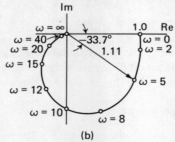

(b)

Fig. 4.6.

The function which is thus plotted is the transfer function with s replaced by $j\omega$. The criterion for stability is the same as for Bode plots, namely, that the output/input magnitude ratio be less than unity when the phase angle is $-180°$, as shown in Fig. 4.7.

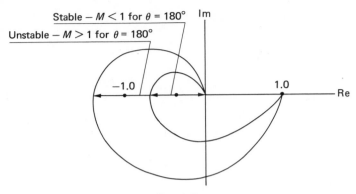

Fig. 4.7.

In Fig. 4.8 the Nyquist plots for several different combinations of elements are sketched so that a number of observations can be made. First, it will be noticed that each single-order element imposes a *maximum* phase lag of $-90°$ on the system of which it may be part, a second-order element $-180°$, so that neither of these two elements can by themselves be unstable. A third-order system can impose more than $-180°$ phase lag and can therefore be unstable as the magnitude ratio can be greater than unity at $-180°$ phase angle.

Dead time or transport lag imposes a fixed time delay input to output without any attenuation. The effect expressed in phase terms is equivalent to

$$\frac{L}{T} \times 360 \equiv \frac{L}{1/\omega} \times 360 \equiv L\omega \times 360° \text{ phase lag} \qquad (4.6)$$

where L is the pure time delay and T is the 'period' of the forcing sinusoid.

Now

$$\frac{\phi(j\omega)}{I(j\omega)} = G(j\omega) \text{ the system transfer function}$$

or

$$\phi(j\omega) = I(j\omega) \times G(j\omega)$$

and

$$I(j\omega) = \Sigma C_n \cdot e^{j\omega t} \text{ for sinusoidal forcing.}$$
$$\therefore \phi(j\omega) = \Sigma C_n e^{j\omega t} \times G(j\omega)$$

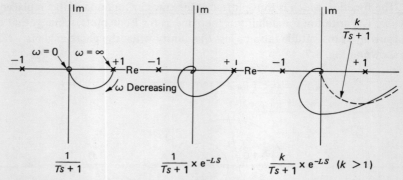

(a) Typical nyquist plots for first-order components showing the effect of pure time delay and gain

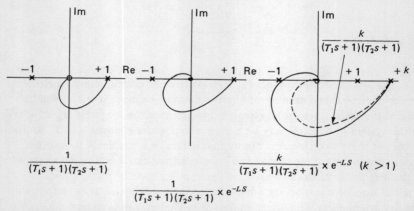

(b) Typical nyquist plots for second-order components showing the effect of pure time delay and gain

Note: (-1) is more likely to be encircled then first-order system

Fig. 4.8.

introducing a pure time delay L, t becomes $(t - L)$ at the outlet: hence

$$\phi(j\omega) = \Sigma\, C_n\, e^{(t-L)j\omega} \times G(j\omega)$$
$$= e^{-Lj\omega} \cdot C_n\, \Sigma\, e^{tj\omega} \times G(j\omega)$$

hence

$$\frac{\phi(j\omega)}{I(j\omega)} = e^{-j\omega L} \times G(j\omega) \qquad (4.7)$$

The transfer function of pure delay is therefore e^{-Ls} (replacing $j\omega$ by s to obtain the general case).

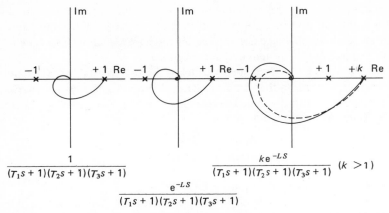

(c) Typical nyquist plots for third-order component showing the effect of pure time delay and gain

Note: (-1) Point is encircled—Hence component described by last diagram is unstable with pure time delay, though not without

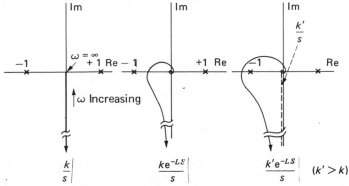

(d) Typical nyquist plots for integration component with and without pure time delay and showing the effect of changes in gain

Note: Pure time delay can make this element unstable whereas without pure time delay it cannot be unstable.

Fig. 4.8.—*continued*.

The pure delay element is the fifth basic element (which should be added to the list given earlier) into which any system can be 'decomposed'.

4.4 Design by the root locus method

Clearly system design must establish the control mechanism transfer function $H(s)$ which will provide satisfactory roots of the system characteristic equation in the sense that no root is actually positive and also that the response, if oscillatory, is acceptably damped for practical purpose. It should be recalled at this point that we are talking here of only one form of control mechanism, feedback, and also of single non-interacting loops. We shall see later how a multiple input/output system may be designed.

In 1950 W. R. Evans developed a search technique to establish by graphical procedures the locus, in the complex plane, of the roots of a system characteristic equation as $H(s)$ varies. The technique rests on the general expression for the transfer function of a system

$$\frac{\phi(s)}{I(s)} = \frac{G(s)}{1 + G(s)H(s)} \qquad (4.8)$$

giving the characteristic equation $1 + G(s)H(s) = 0$.

Since $1 + G(s)H(s)$ is generally a complex quantity it can be expressed in magnitude and phase angle terms

$$G(s)H(s) = -1 \angle n \times 180° \quad \text{where } n \text{ is odd.}$$

If we look once again at the general second-order transfer function in eqn. (4.8) the characteristic equation for which, as was seen in eqn. (2.43) is $s^2 + (a+b)s + (ab + KH) = 0$ where $G = K/(s+a)(s+b)$, we can see that $G(s)H(s)$ which is the 'open loop' transfer function is $KH(s)/(s+a)(s+b)$.

Assuming the transient response of the control mechanism to be instantaneous

$$\frac{KH}{(s+a)(s+b)} = -1 \angle n \times 180° \quad (n \text{ odd}) \qquad (4.9)$$

We have already seen that the value of KH determines whether the closed-loop behaviour is oscillatory, i.e., whether the closed-loop roots are complex or not.

The coordinates of s in the complex plane for any particular value of H must satisfy both the magnitude and angular equality of the closed-loop characteristic equation (4.9). For s to have purely real values, corresponding with non-oscillatory closed loop response, the roots must lie on the real axis. The angular requirement can only be satisfied in such a case if the roots lie between $-a$ and $-b$; the angle of the vector representing the first factor of the characteristic equation $(s+a)(s+b) = 0$ is $\tan^{-1}$ (imaginary component of $s \div$ real component of $s + a$). Since the imaginary component of s is zero the phase angle contributed by each factor respectively is $+180°$ and $-180°$ in all such cases. Points lying to the left of $-a$ or to the right of $-b$ cannot satisfy this requirement, as the total vector angle will then be $\pm 360°$, which does not satisfy the 'n odd' requirement.

As an example let us suppose that the process has the values $a = 3$, $b = 1$. For the closed-loop characteristic equation to have complex roots, we know that the real part of both roots must be the same to give a conjugate pair, and the only such value lies on a vertical line through -2. It can be seen from

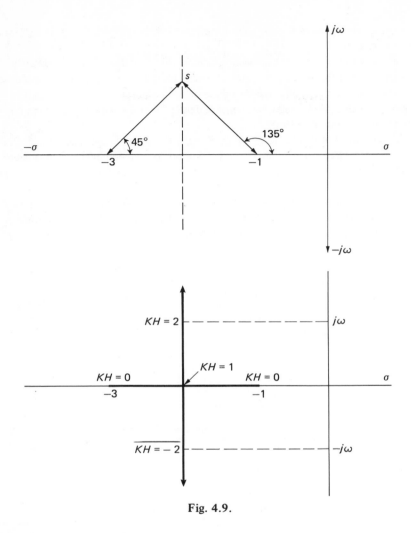

Fig. 4.9.

Fig. 4.9 that any point lying on this line does in fact satisfy the angular requirement, as the phase angles due to each factor of the characteristic function always add up to 180°.

Turning to the magnitude requirement, it will be seen that substituting either -3 or -1 for s in (4.9) gives $KH = 0$ and since K has a definite fixed value, it follows that $H = 0$ for $s = -1$ or -3. Substituting $s = -2$ gives $KH = -1$ and as the values of s move further along the vertical line through -2 in either positive or negative direction, the value of KH increases further.

We now have sufficient information to plot s as H increases and it can be seen from Fig. 4.9 that the locus starts from -1 or -3 when $KH = 0$ (no feedback), the system becoming oscillatory at $KH > 1$. As KH increases further the size of the imaginary component increases—the closed-loop system becomes more oscillatory.

In order to elucidate the reasoning behind this construction, and to verify it the angles for other values of s should be calculated. The construction is always based on the fact that all *possible* values of s must satisfy the angular requirement; the value of H (controller gain) which is the only variable, can then be found for any possible root.

This method of establishing a suitable gain for a simple feedback controlled system is well described in almost any book on servo mechanisms. Since the intention of this book is only to introduce the basic principles of system design it will not be described further here.

To conclude this chapter we will look again at the example with which it began—the vaporizer or boiler. Let us assume the following values and apply the methods so far outlined to the design of what appears to be a simple system

Steam pressure of supply	2.2 kg/cm^2
Boiling point of process material	85 °C
Vapour flowrate required	50 per cent of maximum
Time constant of control valve pneumatic motor-Tv	2.0 s
Dimensions of heating coil	(Fig. 4.10)

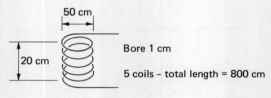

Fig. 4.10.

Latent heat of vaporization of process material at 0 bar is 400 CHV/kg
Latent heat of vaporization of water at 0 bar is 1000 CHV/kg
Heat transfer coefficient 0.010 CHV/°C/cm^2/s. (Measured on internal diameter of coil.)
Control valve maximum capacity 3.75 kg/s.
Control valve character—linear—(equal signal changes/equal capacity change at constant pressure drop). For the sake of simplicity measuring time constants and time constants of heat transfer are considered negligible.

Total heat transfer surface = 800 × π × 1 cm^2 ≅ 2500 cm^2.
Hence maximum heat transfer rate = 25 CHV/°C/s.
Temperature difference steam to process = 50 °C.
Hence maximum heat transfer = 1250 CHV/s.
At 1000 CHV/kg LATENT HEAT ≡ 1.25 kg/s of steam = maximum condensation rate required.

Ratio maximum capacity of control valve to maximum condensation rate is 3.75/1.25 = 3.0/1 = 3.
Considering the response to a sudden increase in vapour flow rate requirement.

Discharge of condensate will be rapid, so that the only time constant of significance (ignoring heat transfer time constants as insignificant) is that of the control valve motor.

Hence $G(s) = \dfrac{3}{(s + \frac{1}{2})}$ and $G(s)H(s) = \dfrac{3H}{(s + \frac{1}{2})} = -1 \angle 180° n$ (n odd)

From angular considerations the single root s must always lie on the real axis to the left of -2 and the root locus plot is therefore as in Fig. 4.11.

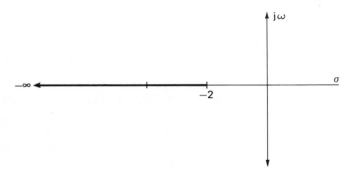

Fig. 4.11.

Considering next the response to a sudden decrease in vapour rate required:

Time constant of heat exchange rate of change is the time taken to change the rate 100 per cent at 100 per cent change of condensate rate, i.e., 1.25 kg/s.

Now total volume of coil $= \pi \times (\frac{1}{2})^2 \times 800$ cm^2
$\simeq 625$ cm^2
and 1.25 kg/s $\equiv 1250$ cm^2/s, hence time constant $= \frac{1}{2}$ s.
The transfer function $G(s)$ of the process, without control is therefore

$$\dfrac{3}{(s + \frac{1}{2})(s + 2)}$$

and the closed-loop characteristic equation

$$\dfrac{3H}{(s + 0.5)(s + 2.0)} = -1 \angle 180° n \text{ (}n\text{ odd)}$$

gives the root-locus diagram Fig. 4.12.

Thus, to avoid oscillatory response altogether with simple proportional (amplifier) type feedback control, the gain must be restricted to $H = 0.187$. This is normally stated as the 'proportional band width' 100/0.187, 535 per cent

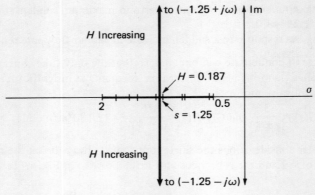

Fig. 4.12.

in this case. The proportional band width (PBW) is the percentage of the controller output signal required to move the final control element—the valve—from one extreme position—closed—to the other—open.

The 'root-locus' diagram shows that for values of controller gains between zero and 0.187 the closed-loop roots are wholly real. For higher gains (narrower proportional bands) the transient response becomes more and more oscillatory. This analysis, however, only applies to the situation when the control action results in the condensate accumulating. As already noted, condensate can be reduced much more quickly and the response is therefore asymmetrical as illustrated in Fig. 4.13 which shows the response to a step load disturbance. If low gain (wide proportional band) is used the response to a disturbance, though non-oscillatory, will be very slow, and, in addition, the steady-state error will be large. If higher gains are used, the integral error will be large.

If the control valve time constant is decreased to $\frac{1}{2}$ s, the two roots of G will be identical and the system will be oscillatory (in the increasing condensate

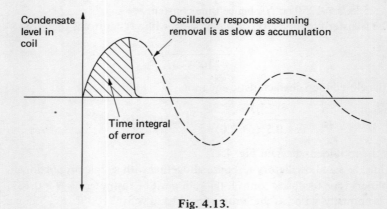

Fig. 4.13.

level case) for all gains greater than zero, as shown by the root-locus diagram shown in Fig. 4.14(a).

If the control valve time constant is even greater, say 8 s, the root-locus diagram is as shown in Fig. 4.14(b), and it can be seen that gains, up to 0.3, give non-oscillatory response. This gain corresponds to a proportional band width of 300 per cent and this result is obviously of no practical significance.

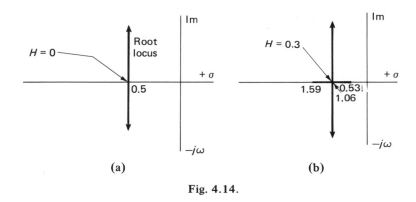

Fig. 4.14.

It can be seen that fairly large excursions in the heat transfer rate and hence the measured value variable are inevitable with such a system unless the speed of response of the control valve is very slow indeed, though, due to the asymmetry of the system, actual oscillation will not occur.

The practical solution to this problem is to change both the condensate accumulation and the control valve time constants. The former can best be achieved by reducing the bore, and thus the volume to heat transfer surface ratio. Adding fins to the tube will further increase heat transfer and thus the rate of accumulation.

It is useful to reconsider this problem with the following variation:

> When the boiling point of the process material is 100 °C. The heating coil is fitted with a steam trap thus enabling condensation to take place at any pressure up to steam pressure.

With these changes the system will no longer 'waterlog' in order to vary the heat transfer surface, for, if the pressure in the coil is atmospheric, the condensing temperature of the steam is the same as the boiling temperature of the process liquid and no heat transfer takes place. Hence, the control valve will be manipulated by the control mechanism, so that maximum inflow rate of steam is achieved at a minimum pressure drop across the valve, of say, 0.35 bar, when the pressure in the coil will be 1.7 bar and the steam will condense at 136 °C, giving a differential of 36 °C for heat transfer into the process liquid. Minimum inflow rate of steam will occur at the full 22 kg/cm² pressure drop across the control valve. Thus, the range of the valve, which is a variable area orifice, must be much greater than before—the low rate opening being approximately the same, but the maximum rate being achieved with a driving pressure drop of 0.35 bar not 2.0 bar. The flow through an orifice being proportional to the square root of the pressure drop, the opening will have to be $\sqrt{30/5} = 2.45$ time larger. Moreover, if we plot the relationship of steam flowrate through the valve against the signal from the control mechanism (assumed to be a simple error amplifier) which was linear in the earlier case, we will find that it now has a greater slope near the maximum flow condition than the minimum flowrate and is closer to a logarithmic or 'equal percentage' relationship. This is a common alternative to linear characteristic offered by control valve manufacturers.

Finally, we might consider the nature of the relating function of this system if—as often happens—the condensed steam has to be lifted 5 m to a collecting main.

5. 'Compensation' by control

5.1 'Proportional' control and error 'reset'

During the course of the discussions of practical systems in chapter 4 certain questions went unanswered; the simple gain or 'proportional' control action in the feedback loop does not answer all control problems adequately. The most obvious deficiency stems from the fact that there has to be an error signal for any corrective signal to be generated. Since the purpose of a control loop is to hold the controlled variable at some value other than its natural steady-state value, this implies that the desired value can never be obtained. In the early days of industrial control, mechanisms were provided with a manual 'reset'—a simple biasing device which provided a false zero for the correction signal at a given set of operating conditions. For instance, in the level control example the equilibrium position of the control valve would be adjusted so that the head in the vessel was at the desired level for the expected input. Under steady conditions no output would be required for the control mechanism when the input was as expected: at any other input value an error was inevitable, its size dependent on the difference of the input from the 'expected' value and the gain. Most modern controllers do not provide any bias mechanism and the size of the error depends on the *full* value of the input.

The need to eliminate offset or error, particularly where high gain cannot be used, as in the case of temperature control, has led to the addition of 'automatic reset' to many control mechanisms. This is achieved by integrating the error with respect to time and adding to the control signal a component proportional to this integral ($180°$ (-1) out of phase). Since the time integral of the error will continue to grow with time, this component will eventually reach the value of the bias required to satisfy the steady-state of the system at any input value. Unfortunately, as was seen in the last chapter, changes of value at one point in a system are not instantly propagated to another due to the time constant elements. The integrating mechanism will be integrating *after* the error has actually disappeared due to the measurement and process time constants and the correction of the process variable will lag behind the increase or decrease of the actual error. The result is that the integral component of the feedback control mechanism tends to cause 'overshoot' or to put it another way it tends to encourage oscillation. For example, take the simple case of the single time constant level process with no significant control mechanism time constants.

$$\frac{\delta h}{\delta I} = \frac{1/C}{s + \left(\dfrac{1+RH}{CR}\right)}$$

Making $H(s) = K + Kk/s$ where K is the gain, k the amplitude factor for the integral term and $1/s$ represents integration, then

$$\frac{\delta h}{\delta I} = \frac{1/C}{s + \dfrac{\left(1 + R\left(K + \dfrac{Kk}{s}\right)\right)}{CR}}$$

$$= \frac{RS}{CR \cdot s^2 + s(1+RK) + Kk} = \frac{1/C \cdot s}{s^2 + \left(\dfrac{1+RK}{CR}\right)s + \dfrac{Kk}{CR}} \qquad (5.1)$$

and the characteristic equation is

$$s^2 + \left(\frac{1+RK}{CR}\right)s + \frac{Kk}{CR} = 0 \qquad (5.2)$$

It can be seen that the addition of an integral term has made the response second-order and that if k becomes large enough oscillatory response is inevitable. Thus, the addition of an integral component always reduces the margin between steady (wholly real) and oscillatory response. The steady-state solution of

$$\frac{\delta I}{C} \cdot s = s^2 \delta h + \left(\frac{1+RK}{CR}\right)s \cdot \delta h + \frac{Kk}{CR}\delta h$$

when $s \to 0$ is $0 = (Kk/CR)\delta h$, and since Kk/CR is constant $\delta h = 0$, that is, there is no error.

Any system is bounded by certain physical limits; for instance, the system resistance can only be varied between an upper and lower limit set by the fully open and fully closed valve positions. Similarly, the integral component generated in a control mechanism has upper and lower limits; yet, in theory, integration will continue as long as there is an error. There is no problem so long as the control signal stays within bounds or limits as the integral function can be negative or positive at any time under these conditions and the definite integral (the value of the bias at a particular point in time) can increase or decrease accordingly. However, the error cannot be removed when control is outside limits, for instance, if the valve has reached an extreme position, and under these conditions the integral component 'saturates'. Its steady-state value increases or decreases until it reaches the physical limit of the control mechanism. As the controller and the control valve are separate items, in many cases the limit set on the control signal may itself be outside or beyond the corresponding limit of the valve which limits actual corrective action. This is the situation when the process is started up.

At start-up the error generates a proportional output which is in the same sense as the integral; thus, if physical limitations are ignored the controller output is the sum of these two components. As the measured value approaches the set point, the proportional component decreases steadily, but since the error has not changed sense (or sign) the integral component remains saturated: indeed, it cannot begin to 'desaturate' until the error does change sense, and 'overshoot' has already occurred (Fig. 5.1(a)). If saturation has continued beyond the normal limit of control action, correcting action will still not take place as the integral component alone exceeds the limit of the control range. Due to the change in error sense, the proportional component now begins to increase, as overshoot continues, in opposition to the integral component which itself decreases for the same reason. Eventually the net output of the controller comes within the control range and the control valve starts to move in the direction to correct the overshoot. Even now overshoot will continue until the corrective action is sufficient to cause the measured variable to start to return toward the set point that it has overshot. The magnitude of the 'initial overshoot' or 'start-up transient' is a function of the proportional band width, or gain, of the controller and of time, since it depends on both proportional and integral actions.

What can be done to minimize the start-up transient? The simplest and most obvious improvement is to limit the integral term so as to prevent it saturating at a value outside the control range; after all, no useful purpose is served by exceeding the limits of control. The integral component will still saturate, but corrective action will now begin as soon as the sense of the error signal changes, that is as soon as the measured value reaches set point. However, overshoot will still occur and on some processes, notably thermal processes with high heat capacitance, it can still be very undesirable. On most applications it is possible to say that during continuous operation the controller output will not exceed certain limits within the control range (say 0.8 bar on a 0.2–1.0 bar range). If the integral component can be limited to this value then, at start-up, corrective action will commence *before* the measured process variable reaches set point (Fig. 5.1(b)). This is best appreciated by considering that at the time when the 'measured value' equals the 'set point' and the output *is* the integral component alone; the control valve has already moved to the 0.8 bar position where previously it would still have been in an extreme position. Fundamentally the integral action of the controller should only come into operation when the measured value approaches the set point on start-up. Because of interaction

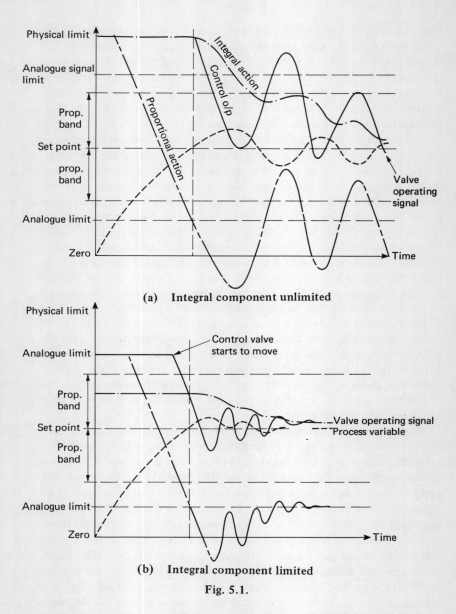

(a) Integral component unlimited

(b) Integral component limited

Fig. 5.1.

between terms this is not practical with pneumatic controllers and was not possible with many early electronic controllers. Today's electronic controllers use highly stable operational amplifiers, and it is quite possible to design such a controller without interaction. Very simple solid state or relay circuitry will enable the integral action to be switched in, at start-up, when the error signal is small and out when the measured process variable goes outside some extreme limit on shut down. Saturation will thus not occur at all and the integral component of the output signal will be zero until it is required for automatic reset during the continuous control period of the batch (Fig. 5.2).

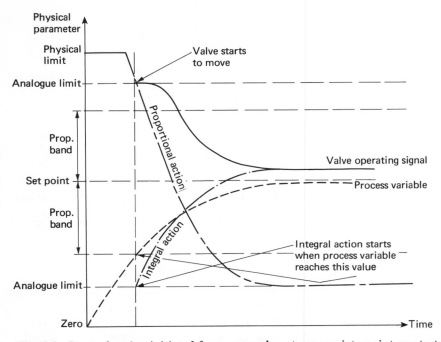

Fig. 5.2. Integral action initiated from zero value at appropriate point on start up.

The addition of an integral component to the control mechanism, then, serves to eliminate significant position error, but has the disadvantages that it makes the system less stable and causes saturation effects at start-up. It is particularly unsatisfactory for processes that start-up and shut-down frequently. It is generally unnecessary with level control systems as the process time constant is normally quite large and the control time constants relatively small, giving essentially single-order response and allowing high gain to be used. Thermal processes, however, usually have large and often multiple measurement time constants and an integral term is almost essential; unfortunately many thermal processes in industry are discontinuous so that 'start-up transients' are often troublesome. Speed or flowrate control systems are liable to suffer from start-up transients because integral saturation causes excessive overshoot, and there are many instances of turbines and compressors which have to be started on hand control as a result. Before leaving the subject of integral control it should be mentioned that it can be used on its own as a feedback control mechanism; thus $H = K_I/s$ where K_I is the integral gain factor. Then

$$\frac{\delta h}{\delta I} = \frac{1/C}{s + \dfrac{1 + RK_I/s}{CR}}$$

$$= \frac{(1/C)s}{s^2 + \dfrac{s}{CR} + \dfrac{K_I}{CR}} \qquad (5.3)$$

and the characteristic equation is:

$$s^2 + \frac{s}{T} + \frac{K_I}{T} = 0 \qquad (5.4)$$

which like the proportional gain + integral action equation is second order and potentially oscillatory.

5.2 'Rate', 'predictive', or 'derivative' compensation

If the addition of an integral term to the simple feedback gain factor gives certain advantages, the addition of a derivative term might reasonably be expected to provide others. If the first derivative or rate of change of the error is measured and the proportional gain factor K increased accordingly, then the control signal applied to the correcting element, the control valve, is proportionally greater than would otherwise be the case whenever the error is increasing or decreasing. The steady-state value of such a component will, of course, be zero—(zero rate of change). The reader will recall that the addition of a 'positioner' to the control valve produces a similar effect and a solution of this problem will now be discussed.

Let

$$G(s) = \frac{s}{C\left(s^2 + \dfrac{s}{RC} + \dfrac{1}{LC}\right)} \qquad \text{eqn. (3.18)}$$

and

$$H(s) = \frac{K + K_d s}{C'/s \left(s^2 + \dfrac{s}{R'C'} + \dfrac{1}{L'C'}\right)}$$

where $K_d s$ is the derivative component of the controller output. Then

$$\frac{G(s)}{1+G(s)H(s)} = \frac{s\left(s^2 + \frac{1}{R'C'}s + \frac{1}{L'C'}\right)}{C\left[\left(s^2 + \frac{s}{RC} + \frac{1}{LC}\right)\left(s^2 + \frac{s}{R'C'} + \frac{1}{L'C'}\right) + \frac{Ks^2 + K_d s^3}{CC'}\right]}$$

which when $C = 0$ gives the characteristic equation

$$s^3 + \left(\frac{1}{T_L} + \frac{1}{T'} + \frac{RK_d}{C'}\right)s^2 + \left(\frac{1}{T'T'_L} + \frac{1}{T_L T'} + \frac{RK}{C'}\right)s + \frac{1}{T_L T'_L T'} = 0 \tag{5.5}$$

Comparing (5.5) with (3.25) and (3.27) it will be seen that if K_d is large it can reduce the negative term under the square root sign in the roots of the closed-loop system characteristic equation. Thus negative derivative action reduces the tendency to oscillation introduced by a positioner.

As might be expected the transient effect of a *positive* derivative component is the reverse of that of the integral component. Hence, while the integral term tends to make oscillation more likely, the addition of a positive derivative term counteracts this in the general case. (The problem of the positioner in the flow control loop is a special case.) Since the steady-state effect is zero we can have our cake *and* eat it with a proportional/integral/derivative 'three term' controller, as far as transient response is concerned. There is, however, no easy solution to integral saturation, and where start-up transients are important integral action should, if possible, be avoided.

One may wonder why the effects of control mechanism inertia, negative derivative, and integral action are not the same when the phase relationship of each is. As has been seen in the valve positioner example, the first two *are* the same; the addition of an integral term, far from curing that problem, exacerbates it. The explanation, mathematically, lies in the particular solution of the differential equation and the effect of constants of integration: the same reasons will explain why, of the three, only integral action removes offset. Both inertia and derivative control depend on rates of change and can have no steady-state value as integral action does.

5.3 Other control actions

One characteristic of both process and control mechanism which has not so far been mentioned is 'dead time' or 'transport lag'. This is caused by slack in mechanisms, 'stiction' or static friction such as is encountered in the gland of a control valve (the reason for using a positioner), or to the actual time taken for the effect of a corrective action to travel from its origin at the control valve or final control element to the detecting element of the measuring device. No amount of phase advance can correct for dead time effects, the error is not even detected until after it occurs. Since the dead time is measured in units of time its effect is worse at high frequencies or in other words in the face of fast input changes. 'Apparent dead time' can arise out of a series arrangement of first-order sub-systems as shown in Fig. 5.3.

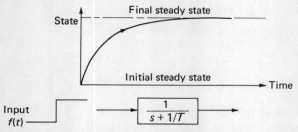

(a) Response of a single time constant element to step change.

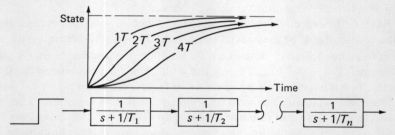

(b) Response of multiple time constant system to step change, when time
(b) constant elements are in series.

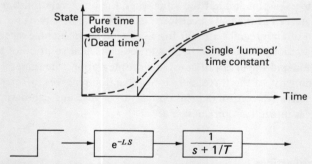

(c) Approximation to transfer function of multiple seriesed time constant elements by a single time constant plus pure time delay (dead time).

Fig. 5.3.

In general nothing can be done to eliminate dead time effects other than to see at the design stage that they do not occur or that if they do they are minimized. The worst of these can be prevented by judicious sampling techniques, but the ability to cope with fast changing inputs will always be impaired by dead time.

To conclude this chapter an interesting case of a process which required negative derivative control action will be given. It concerns the water level control of the steam drum of marine and certain waste heat boilers. Modern drums tend to be relatively small, holding only a fraction of the total boiler contents: a fast change of firing rate consequent on the demands of the steam load controller causes two process responses both of which are detected by the level measuring device. The earliest and fastest of these is the expansion of the water in the boiler which is manifest in the drum, amplified by the relative volume of the drum to the total boiler contents. The second process response is the 'real' one of increased evaporation rate but this is accompanied by a transient pressure increase which tends to suppress the increase of evaporation temporarily. Unfortunately the two effects are in opposition, the unwanted one occurring at a high rate of change and being detected first is soon overtaken by the 'real' response so that the total response to a step change of load would be as shown in Fig. 5.4. It is found, in practice, that the addition of a suitable amplitude negative derivative component to the level control mechanism increases the 'inertial' damping and provides a compromise control response

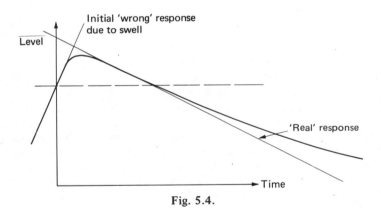

Fig. 5.4.

which largely 'ignores' the initial rapid 'wrong' process response. Such compromise solutions are not uncommon in control system design and a clear understanding of the basic principles is invaluable in arriving at these compromises.

Similar effects occur in distillation columns and, indeed, 'inverse response' is not as rare as might be thought. In a 'column' a step increase in 'boil-up' rate will cause the heavier fraction to be driven further up the column so that the temperature on lower trays increases. Initially, however, the increased vapour flowrate up the column tends to carry away more liquid from each tray, reducing the hold-up of liquid; a transient overflow from the top tray occurs due to this reduced capacity, and the tray below will exhibit a transient of longer duration because it receives the overflow from the tray above. Lower trays will exhibit significant initial 'wrong' response since the transient liquid flowrate increase down the column has the opposite effect to the increase in vapour flowrate which follows.

Chapter 2 introduced the principles of feedback control and subsequent chapters have discussed techniques based on this type of control. Control based on detecting an error which has already occurred has obvious disadvantages, however, particularly when the process exhibits 'dead time'. Unfortunately, many processes do this because the majority of variables are flowrates of materials or energy. However, it is also true that disturbances can often be measured before they enter the process. If we have any knowledge of how the process will respond to the disturbance, it should be possible to reduce its effect by making a compensatory change in the manipulated variable before an error can develop. This is the principle of feedforward control action; this is added to that of the feedback control, which then has a smaller error to deal with. In feedforward control we use our knowledge of the process response to reduce the effect of a disturbance. Because our knowledge can never be perfect there will always be some error, and for this reason feedforward control can never be used alone, but is combined with feedback control.

A simple example of feedforward control is found in the drum level control of large boilers. Increase or decrease of steam flow is measured and a proportional signal added to the output of the water level controller, opening the feedwater valve further before any change of level can occur. The integral term of the level controller eliminates any error resulting from over or under compensation. If a second controller is employed to proportion the feedforward signal a negative derivative term can be added.

Computers can be used to implement more complicated feedforward control schemes. If a mathematical 'model' of the process can be formulated, the computer can calculate the 'future' response to any disturbance detected at input, and formulate a control action which will prevent any response resulting from it. For a process with response similar to that shown in Fig. 5.3(*b*) and (*c*) the advantage is obvious; dead time is overcome. Feedback 'trim' can be added by comparing actual with predicted response at some later time, when any response permitted by imperfections in the model have had time to develop and can be measured. Knowledge of such errors is best employed to correct the model (Fig. 5.5).

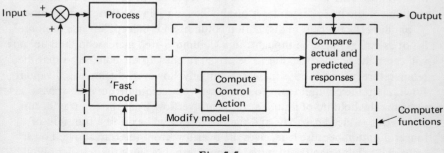

Fig. 5.5.

6. Introduction to multivariable systems

6.1 Models of systems

A process system comprises a number of parts or sub-systems which interact with each other but which are nevertheless capable of operating together in an optimal fashion. 'Optimality' in such a system is achieved by maintaining the essential parameters of the system (those that determine the 'quality' of the product) constant, while other parameters may, in general, vary. The variations can be made to compensate for uncontrollable environmental variations, in such a way that optimality, as defined, is maintained. There will therefore be an optimum value for each of the parameters which determine product quality, and the system will attain its optimum operating condition only when all these values are attained simultaneously. To maintain this optimum state the sub-systems must remain in *dynamic* stability with one another and with the environment—stability being defined in the conventional manner. A process operating in an optimal fashion, therefore, is one in which change in the essential parameters is minimized. Continuous processing requires self-regulation to ensure this minimization. Such regulation can only be achieved by organizing data in such a way as to constitute a 'model' of the system's relationship with its environment, and using this model to determine a suitable control strategy. Whenever the relationship of the system to its environment alters this model must be corrected.

In chapter 1 it was shown that the 'states' of a system are the dimensions of that system in the sense that the system is defined by the behavioural relationships between the physical elements comprising the system. We must choose and define 'states' in a way that suits the 'behaviour' in which we are interested; for instance, if we are concerned with the ability of a boiler to produce steam at a certain pressure and temperature to satisfy certain demands the states we would define would probably be

1. steam pressure p,
2. steam temperature T,
3. water level l.

On the other hand, if we were concerned with the way in which fluctuating boiler pressure will affect the stress level and perhaps cause fatigue failure we would define totally different states because the 'system', although involving the same physical object, the boiler, is totally different. In the first case it is the 'mechanism' of performance, in the second case the mechanism of stress which interests us. Over a period of years we may, in considering the performance of the boiler, have to include, as a relevant consideration, the effect on the stress level of corrosion, so that a broader system combining both sets of elements may have to be defined. Similarly, in considering the control of chemical processes, the chemical engineer will define the mechanism of chemical reaction and select states which are relevant; the control engineer will define the transient behaviour of each of the relevant parameters in terms of instant value, rate of change (velocity) acceleration, etc., that is the dynamic behaviour of each parameter. Significant excursions in, say, temperature for a short period can invalidate design process calculations and spoil the product. Many constraints in the design of plant could be avoided if the dynamics of the process and the chemical mechanisms could be combined into one 'system' and means found to control this broadly defined system. For instance, capacity in a process vessel, necessary to ensure *steady* level or temperature, may actually be detrimental to the optimum mechanism of chemical reaction. By controlling the reaction and the dynamic behaviour of the parameters of that reaction together, improvements can be made in terms of capital expenditure in plant, and quality and yield of product. In order to implement this conception it is necessary to develop techniques for analysing the chemical mechanisms including the dynamic behaviour of the elements of the process system and for synthesizing the control relationships required.

Consider the boiler and separately fired superheater shown diagrammatically in Fig. 6.1. Making the simplifying assumption that the water feed is at saturation temperature, the defining equations can be written

boiler water level $l = a.f(W - Q_s)$
steam pressure $p = b.f(E_b)$
steam temperature $T = c.f(E_s)$

a, b, and c are scalar multipliers—gains.

In this simplified system each of the defining relationships between stimuli (($W - Q_s$), E_b, E_s) and response (l, p, T) are independent of the others; a

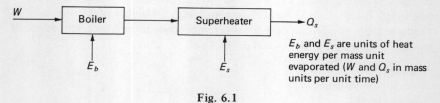

Fig. 6.1

change of level does not affect the boiler pressure, for instance. This fact is reflected in the form of a relating table:

	$(W - Q_s)$	E_b	E_s
l	a	0	0
p	0	b	0
T	0	0	c

which has terms only on the diagonal as shown.

Since, because they are entirely independent, each control loop can be treated as a single input/single output system, the methods already described in this book are quite adequate for the design of such a system. Indeed, such systems are rarely designed at all in the sense that gains and compensation time constants are arrived at during design. The development of the single-loop process analog controller has really been oriented to 'tuning' on site and, provided the engineer who does the tuning has a subjective understanding of the principles involved, it is unnecessary for him to calculate settings which are very easily established by trial and error means.

Trial and error means and the design methods already described both fall short of requirements when several system relationships are other than wholly independent, making it impossible to regard each single input/single output control loop as an *independent* system. Indeed, the established analog computing devices fall short of the hardware requirements at this point. Until the advent of digital computers the only solution to this problem was to make the best compromise possible using single loop controls. The digital computer supplies the hardware required to control a multivariable system with non-independent relationships, and matrix algebra the means to extend the techniques already developed for single input/output systems.

The techniques of representation of multivariable system relationships in matrix form is clearly an essential prerequisite to the adoption of real time computer control if this is to represent an advance in the technology of process control.

6.2 System interaction

The system shown in Fig. 6.2 is typical of a chemical complex.

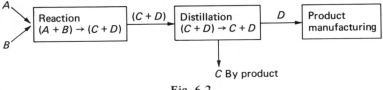

Fig. 6.2

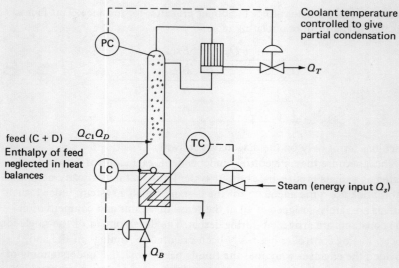

Fig. 6.3.

The distillation system might be as shown in the flow diagram (Fig. 6.3), assuming that the two components of the mixture $(C + D)$ have widely different boiling points. The mechanism of distillation depends on the fact that the composition of vapour in equilibrium with liquid is richer in the more volatile component than is the liquid itself; thus if the vapour formed by heating the liquid (in the column) is removed and condensed in another place (the condenser) the composition of the condensate will be lower in the more volatile component, some of which will therefore remain in gaseous state and can be removed. At any given pressure this disparity between liquid and vapour compositions is greatest at some discrete value of temperature; hence for a given rate of 'energy in' to boil up the liquid in the column, the rate of removal of the more volatile component depends on temperature. If both temperature and pressure can be held steady and the feed composition does not vary, then the boil-up rate can be made a function of the feed rate which, in turn, will be directly proportional to the rate of removal of the more volatile component. These steady 'states' are normally achieved by providing storage between the distillation and the manufacturing systems as shown in Fig. 6.4.

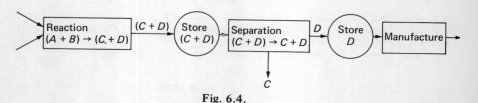

Fig. 6.4.

It is the environment of the distillation process which is being controlled in order that the process itself may proceed 'in steady state'. The input 'buffer store' enables feed rate and composition to be substantially constant, while the output store allows a steady output from the distillation system to match a fluctuating manufacturing rate.

Failure to design a system in which the output of the reaction process matches the input to the distillation process at all times and the output from distillation is accepted as the input to manufacture has to be paid for by provision of two buffer storage vessels. Assuming that the reaction process output cannot readily be made to match the manufacturing process we can see that if the forward flowrate of D can be regulated in the distillation process the second tank can be eliminated.

The problem to be overcome is that none of the states of the distillation system is independent (with the practical exception of column bottoms level control). In fact they are highly dependent. The practical solution shown in the flow diagram (Fig. 6.3) utilizes standard 'single loop' analog controllers, but even though output storage is provided, this solution depends on ensuring that the dynamic response of the temperature loop (which is bound to be 'slower' than the pressure loop) is sufficiently different to reduce coupling effects to a negligible level.

In order to maintain constant distillate and 'bottoms' composition, the temperature must also be constant for the separation of a mixture such as benzene–toluene, for which the liquid/vapour diagram is as shown in Fig. 6.5.

In the example of boiler and superheater each system relationship related only one input and one output, so that the system consisted in effect, of three smaller non-interacting systems. In this example, however, the system relationships include several variables and the system cannot be easily subdivided into

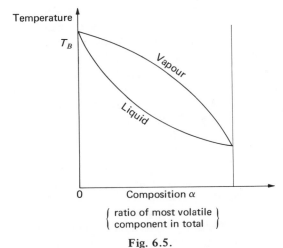

Fig. 6.5.

simpler and independent sub-systems. A change in any of the 'natural' inputs to the system, feed flowrate, feed composition, steam flowrate, etc., may affect not one but all the internal states, level, pressure temperature distillate and top compositions and flowrates, etc.

6.3 Steady-state dimensions

The system states, temperature, overhead flow, etc., represent coordinates of the system space. In this instance, it is a multi-dimensional space. The coordinates, however, are not independent. (Orthogonal in geometrical terms.) We can illustrate what this means in the case of a two-dimensional system as shown in Fig. 6.6.

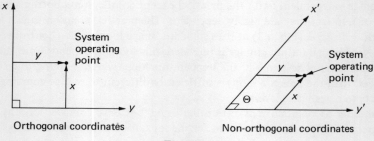

Fig. 6.6.

In the left-hand figure the system operating point is defined in terms of the orthogonal coordinates or states x and y. There is a *direction* in 'space' in which the system operating point can change in one case only affecting x and in the other only affecting y (x and y might represent temperature and pressure, say). In the right-hand figure the operating point is defined in terms of non-orthogonal coordinates and no matter in what direction the system operating point moves both x and y states are affected. Interpreting this in terms of process system states, we must realize that it is no more than a fortunate coincidence if the 'natural' measured states of a process happen to be independent. Indeed, it is very rarely that this is the case; however, often the degree of dependence is small (θ in the right-hand diagram approaches 90° orthogonality) and can be ignored for practical purposes. The control system on p. 90 has been designed by selecting pairs of measured and manipulated variables in such a way as to minimize the interactions within the total system (process plus control). The selection of pairs was made in this case from knowledge and experience of the physical relationships of the process. This may not always produce the optimum system, and for a large system having a great many dimensions may not even be possible. Bristol has shown how the selection can be made in a logical fashion either from the available process design data or from the results of systematic testing on site. An array or table is constructed, having as its columns the inputs to the system (whether manipulable or not) and as its rows the measured variables of the system. Into each space in this table is inserted the 'open-loop gain' between the relevant measured value and the corresponding input. The open-loop gain is defined as the measure of the affect of a change in the input on the measured value, assuming all other variables do not change. Thus:

	u_1	$u_2 \ldots u_n$
x_1	a_{11}	$a_{12} \cdots a_{1n}$
x_2	a_{21}	$a_{22} \cdots a_{2n}$
.	.	. .
.	.	. .
x_n	a_{n1}	$a_{n2} \quad a_{nn}$

is an 'array' which defines the relative gains between the n inputs $u_1, u_2, \ldots, u_n$ and the n measured values $x_1, x_2, \ldots, x_n$, a_{ij} being the gain between u_j and x_i. For convenience, this array can be rearranged in the following manner without changing the meaning in any way. It will be appreciated that this equation represents a set of equations of the form

$$x_n = a_{n1}u_1 + a_{n2}u_2 + \cdots + a_{nn}u_n$$

$$\begin{bmatrix} x_1 \\ x_2 \\ x_n \end{bmatrix} = \begin{bmatrix} a_{11} & a_{12} & \cdots & a_{1n} \\ a_{21} & a_{22} & \cdots & a_{2n} \\ a_{n1} & a_{n2} & \cdots & a_{nn} \end{bmatrix} \begin{bmatrix} u_1 \\ u_2 \\ u_n \end{bmatrix}$$

Note: The columns of x and u terms are referred to as 'vectors' and denoted **x** and **u**.

The open-loop gain of any state x_i to any input u_j represents the *steady-state* change in x_i which will result from a unit change in the input variable u_j in normal operation. This figure will normally be obtainable from a *steady-state* design such as a chemical engineer must produce. If we can define some overall measure of the effect of the input vector **u** on the state vector **x** and then eliminate any particular pair of state/input variables x_i and u_j from these vectors and obtain the corresponding overall measure without them, we can assess the degree of coupling of this pair of variables. If the measure is reduced by exactly the inverse of the steady-state 'open-loop gain' then we know that the two variables under consideration are independent. We can illustrate this by considering the three-dimensional system shown in Fig. 6.7. The states x_1, x_2, and x_3 have been increased by changes in inputs u_1, u_2, u_3, and the overall measure of effect of the change in the vector **u** is the volume change represented by $(x'_1 \cdot x'_2 \cdot x'_3) - (x_1 \cdot x_2 \cdot x_3)$. In the special case where the states x_1, x_2, x_3

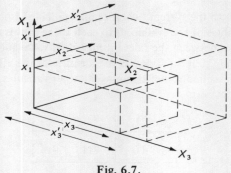

Fig. 6.7.

are all orthogonal, we now consider the two-dimensional sub-system x_1, x_2, ignoring x_3 (and u_3) in Fig. 6.8.

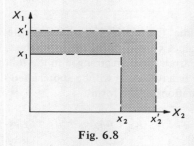

Fig. 6.8

The overall effect of the change in vector **u**, is now the area shaded. If we now multiply this area by $(x_3' - x_3)$ we obtain the total effect again. In matrix terms this two-dimensional subspace is obtained by eliminating the row and column containing the element a_{jk}:

$$\begin{bmatrix} a & 0 & 0 \\ 0 & b & 0 \\ 0 & 0 & c \end{bmatrix} \text{ becomes } \begin{bmatrix} b & 0 \\ 0 & c \end{bmatrix}$$

The same principles can be extended to the non-orthogonal case where the states are dependent. In such a case each u_j has a component effect on each variable x_1, x_2, x_3 and if we eliminate the effect of, say, u_2, on x_2, we only eliminate a component of the total effect of u_2 on the vector **x**. Thus the 'volume' $(x_i x_2 x_3)$ is not reduced by the whole amount represented by the effect of the input u_2.

The measure of overall effect (volume) of the input vector on the state vector is given by the determinant of the matrix (which is literally the volume in the case of a three-dimensional system and area in the case of a two-dimensional system). Hence the relative overall effect of the complete input vector to the reduced subspace is given by

$$\psi_{ij} = \frac{A_{ij}}{|A|}$$

where $|A|$ is the determinant of $[A]$ and A_{ij} is the determinant of the reduced matrix w with input (u_j) and system state (x_i) removed. The matrix whose elements are these factors is known as the relative gain array RGA. If we obtained a matrix such as this

$$\begin{bmatrix} x_1 \\ x_2 \\ x_3 \\ x_4 \end{bmatrix} = \begin{bmatrix} \psi_{11} & \psi_{12} & 0 & 0 \\ \psi_{21} & \psi_{22} & 0 & 0 \\ 0 & 0 & I & 0 \\ 0 & 0 & 0 & I \end{bmatrix} \begin{bmatrix} u_1 \\ u_2 \\ u_3 \\ u_4 \end{bmatrix}$$

for the system under consideration we would know that x_3/u_3 and x_4/u_4 are independent pairs which can be controlled by single-loop controls, thus reducing the size of the interactive system. The matrix this produces can be made more like that which was found to describe the boiler system earlier, by rearranging the pairing of the input and state variables in such a way that the largest terms in the matrix lie on the 'leading' diagonal. In doing this we are minimizing the system interactions as far as is possible. It can now be seen that a control system comprising only single-loops can only implement the diagonal terms of such a matrix of relationships, and ignores the terms which are not on the leading diagonal and which 'describe' the system interactions.

6.4 A simple example

The matrix generated by Bristol's technique constitutes a steady-state model of the system, in that it describes in steady-state terms the influence that each input exerts over *all* the state variables or dimensions of the system. A model of a system can be used to implement multivariable or 'interactive' control as is illustrated by Fig. 6.9. The steady-state behaviour of the system can be described by the matrix

	Q_s/Q_f	Q_r/Q_f	Q_b/Q_f	Q_a/Q_f
Z_d	a_{11}	a_{12}	a_{13}	a_{14}
Z_b	a_{21}	a_{22}	a_{23}	a_{24}
L_b	a_{31}	a_{32}	a_{33}	a_{34}
P	a_{41}	a_{42}	a_{43}	a_{44}

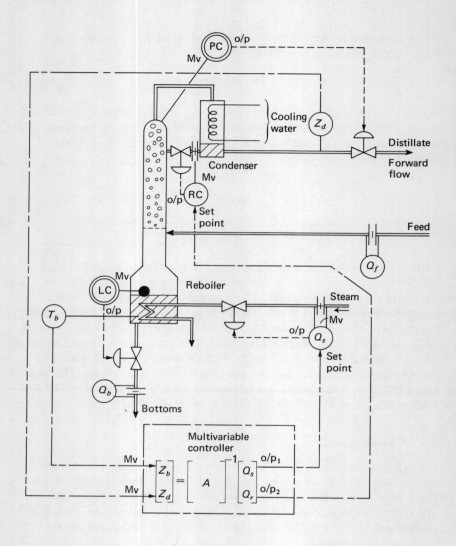

	Q_s	Q_r
Z_b	ϕ_{11}	ϕ_{12}
Z_d	ϕ_{21}	ϕ_{22}

$[A]^{-1} =$

Note: If feed flow rate is not constant Q_s and Q_r can be replaced by Q_s/Q_f and Q_r/Q_f thus 'normalizing' these quantities.

Fig. 6.9.

where Q_s = Steam flow to reboiler
      Q_r = Flow of refluxed distillate
      Q_b = Bottoms product flow
      Q_d = Distillate product flow
      L_b = Reboiler level
      P = Column operating pressure

It would be possible to construct a control system to implement this matrix 'model', but this is not really necessary in practice because the system behaviour is divided into two almost independent parts by reason of the nature of the physical laws which govern it. The state variables will be seen to comprise two compositions and two 'hydraulic' variables (L_b and P). For reasons which are explained in detail in chapter 10, composition variables change very slowly in comparison to hydraulic variables. (All the inputs are flowrates, and have been 'normalized' with respect to feed flowrate Q_f.) The bottom right-hand four elements form a reduced matrix relating Q_b/Q_f and Q_d/Q_f with L_b and P, while the top left-hand four elements form another reduced matrix relating Q_s/Q_f and Q_r/Q_f with Z_d and Z_b. Implementing the first of these two reduced systems closes the mass and thermal balances, while implementing the second closes the composition balance. Consideration of the physical relationships involved reveals that the 'hydraulic' sub-system matrix is of the independent (diagonal) form, for the bottom product flowrate has virtually no effect on operating pressure, while the distillate flowrate has no direct effect on the reboiler level: in consequence this sub-system can be implemented by two independent loops. The regulation afforded by these two control loops can be 'tight' without (by virtue of the differences in response rate) any significant interaction with the composition 'sub-system'* occurring. The interactions between these sub-systems, implicit in the remaining elements of the full matrix (shown shaded) can be resolved by the integral action of the individual controllers. The composition sub-system is not independent, and will have to be implemented in a multivariable manner; however, the variables L_b and P are effectively constant with respect to this sub-system.

The multivariable contribution to the total control scheme has been reduced to the simple equation

$$\begin{bmatrix} Z_b \\ Z_d \end{bmatrix} = \begin{bmatrix} a_{11} & a_{12} \\ a_{21} & a_{22} \end{bmatrix} \times \begin{bmatrix} Q_s/Q_f \\ Q_r/Q_f \end{bmatrix} \quad (6.1)$$

or

$$\mathbf{x} = [A]\mathbf{u}$$

---

* The steady-state 'model' gives us the relative gains of the system, but tells us nothing of the phase relationships. Here the two sub-systems are 'uncoupled' by virtue of a large difference in the frequency components of the composition and hydraulic responses.

where **x** is the vector of system variables and **u** is the input vector. This equation can be re-written

$$\begin{bmatrix} Q_s/Q_f \\ Q_r/Q_f \end{bmatrix} = \begin{bmatrix} a_{11} & a_{12} \\ a_{21} & a_{22} \end{bmatrix}^{-1} \begin{bmatrix} Z_b \\ Z_d \end{bmatrix} \quad (6.2)$$

or

$$\mathbf{u} = [A]^{-1}\mathbf{x}$$

The two compositions comprising the measured value vector **x** can be sampled at regular intervals; at each sampling the input or 'driving' vector can then be calculated according to eqn. (6.2) and the value of the variables Q_s/Q_f and Q_r/Q_f adjusted. If the samples are taken sufficiently frequently, the series of discrete changes in **u** can be made to approximate to a continuous function. Thus, the two input variables are adjusted together in a manner that takes into account the relative magnitude of the affect each has on each of the measured values.

The matrix $[A]$ constitutes a model of the process's relationship with its environment: it is a model, however, only in the steady-state sense. It will be shown in later chapters that the accumulated delays which arise from capacitance within the system and from transport delay result in the affect of Q_s/Q_f on Z_d being out of phase with that on Z_b. This dynamic mismatch may not be serious in a process which has considerable self regulation, but it will be in a process which, without control, shows a tendency to oscillate. For such a process the matrix $[A]$ must be replaced by a model which represents both the steady-state relationships and the dynamic relationships of the process to its environment. How this can be done will be shown in the next three chapters.

If a digital computer is used to implement the eqn. (6.2) it can at the same time be used to compute a conventional three-term control output for each of the single variable loops. For this purpose the computational time of the computer is shared between the various loops; the sampling rate does not have to be the same for each. Thus the complete control scheme can be implemented by the one computer.

7. State space theory

7.1 Dynamic interaction

We noted earlier that the term 'state variable' is used to describe a variable that is the measure of an energy condition within the system. In earlier chapters, it has been amply illustrated that more than one energy storage unit may exist in a single process parameter control loop. The oscillatory and even unstable response of which such a loop is capable has been analysed at some length, as has the means of compensating for it. Responses associated with single process parameter loops are generally of a different order of frequency than those associated with the interactions of the process variables themselves. If we assume now that every process parameter which is a variable or dimension of the system is compensated by a conventional three-term controller, then only process parameters need be defined as the 'state variables' of the system.

The matrix developed in the last chapter described the transition from one steady-state to another, ignoring the 'transitory' behaviour altogether. The dimensions of the system under this steady-state definition are those process parameters which fully define any steady-state condition. Similarly, the dimensions of the system under this dynamic definition are those process parameters by which the state of the system *at any instant of time* can be fully described. As was shown in chapter 1 there are, in general, more dynamic dimensions than steady-state dimensions.

The state equation

$$\frac{d}{dt}\mathbf{x} = [A]\mathbf{x} \qquad (7.1)$$

is first-order and the terms in the left-hand vector are the first differentials of those in the right-hand state vector. Rarely, if ever, will higher differentials (acceleration, etc.) of process variables be as significant as they are when considering control of one of these parameters itself (almost all process parameters are flowrates of material), and it can be seen, therefore, that the elements of matrix $[A]$ will always be real. The matrix notation represents a set of n equations as it did in the previous chapter, but in this case they are first-order differential equations of the form

$$\frac{dx}{dt} = a_1 x_1 + a_2 x_2 + \cdots + a_n x_n$$

The representation of the system behaviour in this manner allows us to manipulate the complete system as easily as we previously manipulated the equations for a single variable. There are, however, more rules to be observed, as one would expect. The columns of system variables and inputs are known as 'vectors' and their elements are the coordinates of the process operating point or the input function as seen in the last chapter. The mathematics of matrices is today very extensive, and represents a very powerful method. For the purposes of this book, however, it is not necessary for the reader to become deeply involved in this branch of mathematics, but before proceeding to consider the problem of controlling the process dynamically it is necessary to become familiar with some of the basic rules.

Since a complete solution for n variables can only be obtained if there are n defining equations, real systems will always be represented by 'square' matrices. These matrices can, as we have already seen, be 'partitioned' in order to consider some part of the system, but that part will always be 'square' also.

The matrix in a matrix equation takes the place of the arithmetical coefficient in a single equation: we have already seen that a vector has directional properties as well as scale, unlike a single algebraic variable. A matrix therefore must in general modify the direction and scale of a vector on which it operates. The property of the matrix to modify scale is expressed as the 'determinant' of the matrix as already noted in the previous chapter. The determinant is associated with the overall gain of the system while the matrix's property to change direction in 'space' is associated with interactive affects of inputs on outputs. The interactive effects are steady-state if the defining equations are purely algebraic as in the last chapter and 'complex' if the system is defined by differential equations.

In the pages that follow the reader will find that matrices are multiplied by other matrices. The need to do this arises out of the combining of the modifications of two sub-systems in series and is analogous to combining two transfer functions in a single parameter system. Since the procedure amounts to n vector matrix operations, the vectors being the columns of the second matrix, it is important not to reverse the order of the matrices involved, since a totally different result will obviously then be obtained. Inverting a matrix is a perfectly valid operation, (see appendix I for details of the calculation of the determinant and the inverse of a matrix). Arising out of the order rule for multiplication is a rule that requires that the inverse of the product of two matrices is given by the product of the inverses of the individual matrices in reverse order.

There are many excellent books available on the subject of matrix algebra (or linear mathematics) for the reader who wishes to study advanced control theory. The purpose of this book is to present the principles as they affect system design, and for the present purpose no further mathematical knowledge is necessary.

We have noted that a system can display oscillatory behaviour if *either* the *process* relationships are of order higher than one, *or* if more than one time constant within the system are coupled together by system interactions. Since the matrix equation expressing the system behaviour is always first-order, we know intuitively that the 'transition' matrix $[A]$ must have some equivalence to the transfer functions developed in earlier chapters for single input/output systems. This relationship is not hard to find: consider the second-order homogeneous expression in matrix form

$$\ddot{x} + \frac{a_1}{a_2}\dot{x} + \frac{a_0}{a_2}x = 0*$$

or in matrix form

$$\begin{bmatrix}\dot{x}_1\\ \dot{x}_2\end{bmatrix} = \begin{bmatrix}0 & +1\\ -\frac{a_0}{a_2} & -\frac{a_1}{a_2}\end{bmatrix}\begin{bmatrix}x_1\\ x_2\end{bmatrix} \qquad (7.2)$$

Alternatively

$$\ddot{x} + 2\zeta\omega_n\dot{x} + \omega_n^2 x = 0$$

or

$$\begin{bmatrix}\dot{x}_1\\ \dot{x}_2\end{bmatrix} = \begin{bmatrix}0 & +1\\ -\omega_n^2 & -2\zeta\omega_n\end{bmatrix}\begin{bmatrix}x_1\\ x_2\end{bmatrix} \qquad (7.3)$$

Adding stimuli or forcing functions to the second-order equation to represent disturbing influences of first-order

$$\ddot{x} + a_1\dot{x} + a_0 x = p_0 u + p_1 \dot{u}$$

the transfer function for which is of the form

$$\frac{p_0 + p_1 s}{s^2 + a_1 s + a_0} \qquad (7.4)$$

This can be represented two ways in matrix form thus

$$U(s) \longrightarrow \boxed{\frac{p_0 + p_1 s}{s^2 + a_1 s + a_0}} \longrightarrow Y(s)$$

$$\ddot{y} + a_1\dot{y} + a_0 y = p_0 u + p_1 \dot{u}$$

$$x_1 = y$$

$$x_2 = \dot{y} - p_1 u = \dot{x}_1 - p_1 u$$

hence:

$$\begin{vmatrix}\dot{x}_1\\ \dot{x}_2\end{vmatrix} = \begin{vmatrix}0 & x_2\\ -a_0 x_1 & -a_1 x_2\end{vmatrix} + \begin{vmatrix}p_1 u\\ a_1 p_1 u + p_0 u\end{vmatrix}$$

or

$$\begin{bmatrix}\dot{x}_1\\ \dot{x}_2\end{bmatrix} = \begin{bmatrix}0 & +1\\ -a_0 & -a_1\end{bmatrix}\begin{bmatrix}x_1\\ x_2\end{bmatrix} + \begin{bmatrix}p_1 + 0\\ a_1 p_1 + p_0\end{bmatrix} u \qquad (7.5)$$

* $\dot{x}$ is used to represent $(d/dt)x$ and $\ddot{x}$ to represent $(d^2/dt^2)x$ the second differential of x.

```
  U(s)  ┌─────────────┐  Z(s)  ┌─────────┐  Y(s)
 ──────▶│      1      │───────▶│ p₀+p₁s  │──────▶
        │ s²+a₁s+a₀  │        └─────────┘
        └─────────────┘
```

$$\ddot{y} + a_1\dot{y} + a_0 y = p_0 u + p_1 \dot{u}$$

$$z_1 = z$$

$$z_2 = \dot{z}$$

hence:

$$\ddot{z} + a_1 \dot{z} + a_0 z = u \text{ and } y = p_0 z + p_1 \dot{z}$$

so that:

$$\begin{bmatrix} \dot{z}_1 \\ \dot{z}_2 \end{bmatrix} = \begin{bmatrix} 0 & 1 \\ -a_0 & -a_1 \end{bmatrix} \begin{bmatrix} z_1 \\ z_2 \end{bmatrix} + \begin{bmatrix} 0 \\ 1 \end{bmatrix} u$$

and

$$[y_1 + y_2] = [p_0 + p_1] \begin{bmatrix} z_1 \\ z_2 \end{bmatrix} \tag{7.6}$$

7.2 Closed-loop stability

In establishing the stability, or otherwise, of a single input/output system, we sought the roots of the characteristic equation. We now want to establish the stability criterion for the multivariable system under consideration: what then in matrix terminology is equivalent to the roots of the characteristic equation?

When a vector is operated on by a matrix the result is another vector which, in general, differs from the first in scale and direction in 'space'. In the special case when the two vectors have the same direction, the operation of the matrix changes only the scale and is therefore equivalent to a scalar operation. An n-dimensional system has n states or variables which are vector coordinates of the 'space' in which that system is defined. If we can find n other vectors *such that for each the operation of the system matrix is scalar*, then each of these vectors represents a combination of system variables *which is independent of all other* such combinations. Defining the system in terms of these vectors rather than the original 'natural' states or coordinates, we will have exactly the same type of system as the boiler and superheater system considered in the last chapter. The n roots of the n independent single-order equations which will thus be discovered are the roots of the system. In general, they may be real or complex and we know that providing none of them has a positive real part the system must be stable.

Can we find n such vectors? If such a vector exists, then,

$$[A]\theta = \lambda\theta$$

hence
$$([A] - \lambda[I])\theta = 0$$

and
$$([A] - \lambda[I]) = 0$$

where θ is the vector and λ a scalar. $[I]$ is the identity matrix for non-trivial cases $\theta \neq 0$.

For a second-order system:
$$[A] = \begin{bmatrix} 0 & +1 \\ -\omega_n^2 & -(2\zeta\omega_n) \end{bmatrix}$$

hence
$$([A] - \lambda[I]) = \begin{bmatrix} (0-\lambda) & +1 \\ -\omega_n^2 & -(2\zeta\omega_n + \lambda) \end{bmatrix} = 0 \tag{7.7}$$

The determinant of this matrix is
$$\lambda^2 + 2\zeta\omega_n\lambda + \omega_n^2 = 0 \tag{7.8}$$

the characteristic function and the scaler quantity λ is clearly given by one of the roots of the characteristic equation.

In this case the matrix is (2 × 2), the system two-dimensional and the two roots or 'eigenvalues' are
$$\lambda_1 = -\zeta\omega_n + \omega_n\sqrt{(\zeta^2 - 1)}$$
$$\lambda_2 = -\zeta\omega_n - \omega_n\sqrt{(\zeta^2 - 1)}$$

The vector θ_1 corresponding to λ_1 is called an eigenvector. Only the direction of this vector need be found for the operation of $[A]$ on any vector of the same direction will be scalar. Hence only θ_2/θ_1 need be found.

Solving the equation:
$$([A] - \lambda[I])\theta = 0$$

or
$$\begin{bmatrix} (0-\lambda) & +1 \\ -\omega_n^2 & -(2\zeta\omega_n - \lambda) \end{bmatrix} \begin{bmatrix} \theta_1 \\ \theta_2 \end{bmatrix} = 0$$

or
$$-\lambda\theta_1 + \theta_2 = 0 \tag{7.9}$$

and
$$-\omega_n^2\theta_1 - (2\zeta\omega_n - \lambda)\theta_2 = 0 \tag{7.10}$$

For θ_2/θ_1 either (7.9) or (7.10) gives

$$\frac{\theta_2}{\theta_1} = \lambda$$

hence, if

$$\theta_1 = 1 \quad \theta_2 = \lambda$$

and the two eigenvectors are

$$\begin{bmatrix} 1 \\ \lambda_1 \end{bmatrix} \text{ and } \begin{bmatrix} 1 \\ \lambda_2 \end{bmatrix} \qquad (7.11)$$

Consider the equation $\dot{\mathbf{x}}(t) = [A]\mathbf{x}(t)$ which describes the behaviour of the closed system without disturbances. We can redefine the vector of system states $\mathbf{x}(t)$ in terms of the new set of independent coordinates

$$\mathbf{x}(t) = C_1 \theta_1(t) + C_2 \theta_2(t) + \cdots + C_n \theta_n(t) = [W] \qquad (7.12)$$

where $C_1, C_2, C_3, \ldots$, are constants indicating the scalar distance along each coordinate of the vector space.

Now $\dot{\theta}_n(t) = \lambda_n \theta_n(t)$ since λ_n replaces $[A]$ when the vector is the nth eigenvector, hence $\dot{\mathbf{x}}(t)$ can be re-expressed thus,

$$\mathbf{x}(t) = C_1 \lambda_1 \theta_1(t) + C_2 \lambda_2\,_2(t) + \cdots + C_n \lambda_n \theta_n(t) \qquad (7.13)$$

or

$$\begin{bmatrix} C_1 \theta_{11}(t) + C_2 \theta_{12}(t) & \cdots & C_n \theta_{1n}(t) \\ C_1 \theta_{21}(t) + C_2 \theta_{22}(t) & & C_n \theta_{2n}(t) \\ \vdots & & \vdots \\ C_1 \theta_{n1}(t) + C_2 \theta_{n2}(t) & \cdots & C_n \theta_{nn}(t) \end{bmatrix} \begin{bmatrix} \lambda_1 & & 0 \\ & \lambda_2 & \\ & & \ddots \\ 0 & & \lambda_n \end{bmatrix} \qquad (7.14)$$

which can be abbreviated to

$$[W][\Lambda]$$

hence $\dot{\mathbf{x}}(t) = [A]\mathbf{x}(t)$ can be re-expressed in the form

$$[W][\Lambda] = [A][W]$$

from which

$$[A] = [W][\Lambda][W]^{-1}$$

and hence

$$\dot{\mathbf{x}}(t) = [W][\Lambda][W]^{-1}\mathbf{x}(t)$$

or

$$[W]^{-1}\dot{x}(t) = [\Lambda][W]^{-1}x(t) \qquad (7.15)$$

defining a new state vector in which each of the new states is some combination of the original 'natural' states $x_1, x_2, \ldots$

$$Z(t) = [W]^{-1}x(t)$$

hence

$$\dot{Z}(t) = [\Lambda]Z(t) \qquad (7.16)$$

and since $[\Lambda]$ is a diagonal square matrix the system so defined is composed of n independent relationships as in the case of the boiler/superheater. The eigenvalues $\lambda_1, \lambda_2, \ldots, \lambda_n$ are the roots of the n independent *first-order* equations defining the system.

Hence

$$z_n(t) = e^{\lambda_n t} \, z_n(0) \qquad (7.17)$$

We know that if any of the solutions to these n equations is a growth term, that is, λ_n has a positive real part, the whole system is unstable, since each element ζ_n of the new state vector $Z(t)$ is in reality a combination of the natural states $x_1, x_2, \ldots, x_n$.

We are now in a position to assess the stability of the system, including those controls which already form part of it (in the example of a simple separation the pressure and bottoms level were independently controlled) to establish whether it displays inherent stability. We also wish to assess the stability of the system with modified or added control relationships and thus determine either:

1. The optimum controls which will restore the desired steady-state conditions after disturbance without unacceptable oscillatory response.

or

2. The necessary controls to make an unstable system, or one which has an unacceptable response, stable.

To achieve this end we must consider the equation

$$\dot{x}(t) = [A]x(t) + [B]u(t) \qquad (7.18)$$

where $[B]u(t)$ represents the inputs to be derived from feedbacks (we ignore the possibilities of feedforward control at present). Since $u(t)$ is a function of the system states $x(t)$ in general

$$u(t) = [C]x(t) \qquad (7.19)$$

and we can re-write the system equation

$$[W]\dot{Z}(t) = [A][W]Z(t) + [B][C][W]Z(t)$$

or
$$\dot{Z}(t) = [\Lambda]Z(t) + [W]^{-1}[B][C][W]Z(t) \qquad (7.20)$$

Combining the two terms on the right

$$\dot{Z}(t) = [K]Z(t)$$

where $[K] = [\Lambda] + [D]$ and $[D] = [W]^{-1}[B][C][W]$. That is

$$\begin{bmatrix} k_1 & & & 0 \\ & k_2 & & \\ & & \cdot & \\ & & & \cdot \\ 0 & & & k_n \end{bmatrix} = \begin{bmatrix} \lambda_1 & & & 0 \\ & \lambda_2 & & \\ & & \cdot & \\ & & & \cdot \\ 0 & & & \lambda_n \end{bmatrix} + \begin{bmatrix} \delta_1 & & & 0 \\ & \delta_2 & & \\ & & \cdot & \\ & & & \cdot \\ 0 & & & \delta_n \end{bmatrix} \qquad (7.21)$$

The roots of the system with controls added are $k_1, k_2, \ldots, k_n$ and it is now possible to choose $\delta_1, \delta_2, \ldots, \delta_n$ so as to modify the original roots $\lambda_1, \lambda_2, \ldots, \lambda_n$ as necessary. If the original system is unstable or has too oscillatory a response, it may be possible to modify some of the components, such as a control valve time constant, in order to implement the required change. As was pointed out earlier, however, if the oscillation or instability arise from process reaction relationships it may prove very difficult to make such alterations. In such a case some modification of the control relationships may be necessary.

7.3 Controllability

There is a correspondence between the matrix representing a multi-state non-independent system and the transfer function of that system with multiple inputs and outputs. It makes no difference whether the 'states' of the system are pressure, temperature rate of flow, etc., or the rate of change, acceleration, or higher differential of any one of these process parameters. To a computer a system state is a mathematical quantity which bears some relationship to another or other states. The correspondence of matrix to transfer functions can be clearly seen in the block diagram representations in Fig. 7.1.

$$\begin{bmatrix} \dot{x}_1 \\ \dot{x}_2 \\ \dot{x}_3 \end{bmatrix} = \begin{bmatrix} a & 0 & 0 \\ 0 & b & 0 \\ 0 & 0 & c \end{bmatrix} \begin{bmatrix} x_1 \\ x_2 \\ x_3 \end{bmatrix} + \begin{bmatrix} l \\ m \\ n \end{bmatrix} u$$

$$\begin{bmatrix} \dot{x}_1 \\ \dot{x}_2 \\ \dot{x}_3 \end{bmatrix} = \begin{bmatrix} a & 0 & f \\ g & b & d \\ c & 0 & c \end{bmatrix} \begin{bmatrix} x_1 \\ x_2 \\ x_3 \end{bmatrix} + \begin{bmatrix} l \\ 0 \\ 0 \end{bmatrix} u$$

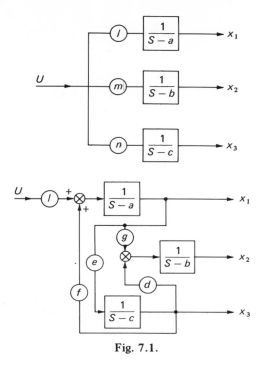

Fig. 7.1.

The first of these two systems consists of three independent first-order sub-systems each with a single input *lu, lm, ln* and a single state variable x_1, x_2, or x_3. If *l, m,* or *n* is zero then one sub-system is disconnected (unaffected by input). The second system is not independent and even though there is only one input $l \times u$ all three states x_1, x_2, and x_3 are affected by any change of input. That the manner of that effect is determined by the matrix is obvious from the block diagram equivalent. It should be noted that the *only* elements comprising the block diagram are simple multipliers and first-order transfer functions. If we trace out the effect of a change of *u* on x_2, however, the path runs through three first-order blocks indicating that the relationship of x_3 to input *u* within the system is third-order.

If a system has more than one controlled variable we need to be able to determine whether they can be simultaneously controlled. This can be illustrated by reference to the example given at the beginning of chapter 6.

Let **x** be the vector of system states

$$\mathbf{x} = \begin{bmatrix} l \\ p \\ T \end{bmatrix}$$

Let $[A]$ be the system relating matrix

$$[A] = \begin{bmatrix} a & 0 & 0 \\ 0 & b & 0 \\ 0 & 0 & c \end{bmatrix}$$

Let **u** be the input vector

$$\mathbf{u} = \begin{bmatrix} w - q \\ E_b \\ E_s \end{bmatrix}$$

Any system state say $\mathbf{x}(t)$ must be attainable by the net effect of each input applied over a finite period of time starting from another state $\mathbf{x}(0)$

$$\mathbf{u} = \int_0^t \begin{bmatrix} w - q \\ E_b \\ E_s \end{bmatrix} dt$$

and similarly $\mathbf{x}(0)$ must be attainable starting from $\mathbf{x}(t)$ by applying the negative total input to each independent sub-system. This is so if the columns of matrix $[A]$ are independent, i.e., if

$$\begin{bmatrix} a & 0 & 0 \\ 0 & b & 0 \\ 0 & 0 & c \end{bmatrix}$$

Suppose, however, that the columns are not independent.

Let $[A]$ be

$$\begin{bmatrix} a & 2a & 0 \\ b & 2b & 0 \\ 0 & 0 & c \end{bmatrix}$$

so that the second column is a multiple of the first making these columns dependent. It can be seen that the effect of *any* set of inputs on level l and 'pressure' p is always in the ratio 2:1. We cannot now define a set of inputs which will drive the system from any given state to any other given state: for instance, we cannot drive the system to a new state in which l remains unaltered but p changes, for if p changes so must l.

8. Multivariable compensation

8.1 Decoupling and control

In the early days of process control, the human operator manipulated the valves and dampers that controlled the flow of materials and energy into and out of the process. Today automation has reduced the amount of manual work required, but it is still men who operate the process as a whole. Regulation of a single parameter is a single dimensional problem, while regulation of the process is a multi-dimensional one. When the operator changes the desired value of one of the parameters of an interactive process, he affects all the other parameters, and these influences vary in magnitude and in the speed at which they occur. Much research has been carried out to find ways in which the process can be 'decoupled', and it is strange that more progress has not been made towards an 'automatic operator'. In order to compensate for process interactions a model is, as we have seen, required; such a model exists in the knowledge and experience of the human operator, but it is at best inaccurate and ill-defined, and changes, moreover, with each change of shift. An automatic operator would not make the human one redundant, but rather would elevate him to a position in which his function would be the management of the process and not merely its manipulation.

One approach to the problem of synthesizing an automatic control scheme for an interacting process is to 'decouple' it and then apply the conventional single parameter loops to the 'uncoupled' process.

The internal sub-systems of the process may interact in such a way that the response to certain types of disturbance exhibits such phenomenon as inverse response, etc. The decoupling device may be very difficult to design, and in many cases, it may demand components which are not possible. It is not always necessary to design a decoupler to perform the whole function of decoupling; the single parameter controller can partially implement it, and it is tuneable on site. Rosenbrock has developed a particularly powerful technique for designing systems in this way; in effect each element of Bristol's relative gain array is replaced by a Nyquist diagram representing the *dynamic* relationship of *one* input to *one* measured value for the special case when all interactions are considered to be suppressed. It is possible to synthesize these Nyquist diagrams from a knowledge of the time constants, transport delays, etc., which would normally be obtainable from process design data. From this 'inverse Nyquist array' a network or device can be designed which 'accomplishes a modification of the interaction in the plant' such that the decoupling can be completed by the dynamic compensation afforded by the independent three-term controllers. This concept is illustrated in Fig. 8.1. It should be appreciated that the alternative control mechanism shown on p. 36 has been assumed here.

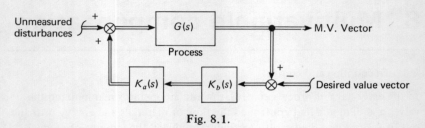

Fig. 8.1.

While this technique is capable of producing a very sophisticated control system for almost any process, it necessitates the detailed analysis and synthesis of each individual system because of the division of the decoupling function.

Figure 8.1 has a closed-loop transfer function

$$([I] - [G]_{(s)}[K_a]_{(s)} \cdot [K_b]_{(s)})^{-1}[G]_{(s)} \qquad (8.1)$$

and can be redrawn thus as in Fig. 8.2, where $[P]_{(s)}$ represents the transfer

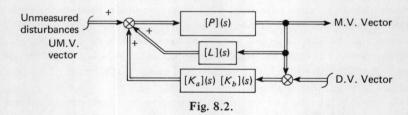

Fig. 8.2.

function of the decoupled process and $[L]_{(s)}$ represents the internal feedback within the process. Assuming the desired values to be constant, the diagram reduces to that shown in Fig. 8.3. where $[H]_{(s)}$ now represents

$$[K_a]_{(s)}[K_b]_{(s)} + [L]_{(s)} \qquad (8.2)$$

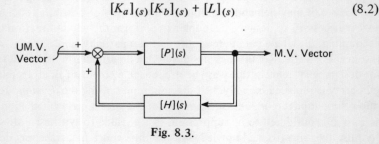

Fig. 8.3.

The closed-loop function of Fig. 8.3 is

$$([I] - [P]_{(s)}[H]_{(s)})^{-1}[P]_{(s)} \quad (8.3)$$

where $[I]$ is the identity matrix. Letting

$$[H]_{(s)} = [P]_{(s)}^{-1}[Q] \quad (8.4)$$

where $[Q]$ is a scalar multiple of the identity matrix, and substituting (8.4) in (8.3) the closed-loop transfer function is

$$([I] - [P]_{(s)}[P]_{(s)}^{-1}[Q])^{-1}[P]_{(s)}$$
$$= ([I] - [Q])^{-1}[P]_{(s)} \quad (8.5)$$

From (8.2)

$$[K_a]_{(s)}[K_b]_{(s)} = [H]_{(s)} - [L]_{(s)}$$

and from (8.4)

$$[H]_{(s)} - [L]_{(s)} = [P]_{(s)}^{-1}[Q] - [L]_{(s)}$$
$$= [P]_{(s)}^{-1}([Q] - [P]_{(s)}[L]_{(s)}) \quad (8.6)$$

But by definition

$$[G]_{(s)}^{-1} = \{([I] - [P]_{(s)}[L]_{(s)})^{-1}[P]_{(s)}\}^{-1} = [P]_{(s)}^{-1}([I] - [P]_{(s)}[L]_{(s)})$$

hence

$$[K_a]_{(s)}[K_b]_{(s)} = [G]_{(s)}^{-1} + [P]_{(s)}^{-1}([Q] - [I]) \quad (8.7)$$

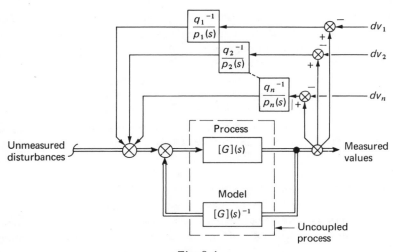

Fig. 8.4.

In the special case when the closed-loop feedback gains are all unity, ($[Q] = [I]$) ideal decoupling is achieved when the decoupling network alone has the transfer function which is the inverse of the process transfer function. Combined decoupling and control is given by eqn. (8.7), the block diagram for which is shown in Fig. 8.4. Since each of the n independent control loops is 'closed' about the decoupled process, it can be seen that a change of any desired value does not interact on other loops.

It is significant that interactive control and feedforward control both demand a model of the system: if the measured value vector in Fig. 8.4 is included input variables which are measurable, though they cannot be manipulated, then the system is seen to combine feedforward and feedback control.

8.2 Sampling techniques

Implementation of the technique suggested in Fig. 8.4 implies that a model of the process can be synthesized to represent, when inverted, $[G]_{(s)}^{-1}$. It has already been said that this is rarely if ever possible, but it will be recalled from chapter 6 that often only certain parts of the system response need be considered in practice. Provided that unwanted responses can be filtered out of the measurement vector, a restricted form of $G_{(s)}^{-1}$ can be used for decoupling purposes. In real systems it is normally the low frequency responses which give rise to difficult interactions and the application of the basic sampling theorem to the selection of the process (not the computer) sampling interval will provide an effective filter, as will now be shown.

The curve in Fig. 8.5(a), (b), and (c) represents the changing value of the process variable over some finite period of time. It can be seen that, if the sample period is shorter than the period of variation of the process, assumed here to be varying in a sinusoidal fashion, and each value sampled is 'held' throughout the following interval, the resultant 'discrete function' is a reasonable approximation to the true function. As the sample period is shortened in relation to the periodicity of the process variable this approximation improves correspondingly, as shown in Fig. 8.5(b).

In the limit, when the sampling period becomes infinitely short, the two functions are identical. Conversely, as the sampling interval lengthens, the approximation of discrete function to true function deteriorates. The question which inevitably arises is 'at what point does the discrete approximation cease to adequately represent the true function'? This is really the wrong question to ask, as it neglects the fact that the function is a sinusoid. Given this extra information it can be seen that, if at least two samples are obtained within the period of the sinusoid, these will provide all the information necessary to fully define the function. (Try to draw more than one sinusoid through two given points lying within one cycle if you find this hard to believe.) Of course, a real process variable will not obligingly behave in a purely sinusoidal fashion, but we know that any function can be broken down into its component fre-

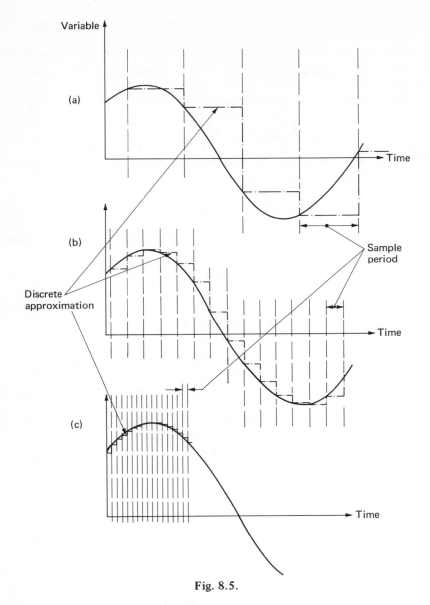

Fig. 8.5.

quencies, and so it can be seen that, provided the sample rate exceeds twice the frequency of the *highest significant* component of the system response, the function will, indeed, be fully defined. (The formal proof of this fact is beyond the scope of this book.)

If the sample rate is selected to faithfully reproduce a portion of the frequency spectrum of the process function and exclude the remainder, the incomplete information obtained for that portion of the spectrum lying outside the 'window' of the filtering effect constitutes random 'noise' in the measurement. Typical process sampling rates will be 0.2–2.0 samples per minute: the computer will probably be capable of at least 100 samples per second on the other hand. By sampling at the higher rate and averaging over each *process* sample interval this random noise can be removed.

8.3 A multivariable algorithm

Provided the appropriate process variables can be measured, the system states can now be defined in such a way that only rates of change of process variables appear in the left-hand vector of the state equation

$$\dot{x} = [A]x \tag{8.8}$$

for which $x = e^{-[A]t}$ is a solution if the exponential function of $[A]$ has a real meaning. Then

$$\dot{x}(t) = [A]x(t) + [B]u(t) \tag{8.9}$$

hence

$$[A]\,e^{[A]t}Z(t) + e^{[A]t}\dot{Z}(t) = [A]\,e^{[A]t}Z(t) + [B]u(t)$$

hence

$$\dot{Z}(t) = e^{-[A]t}[B]u(t)$$

Integrating

$$Z(\tau) - Z(0) = \int_0^\tau e^{-[A]t}[B]u\,dt$$

hence

$$Z(\tau) = Z(0) + \int_0^\tau e^{-[A]t}[B]u\,dt \tag{8.10}$$

where $Z(0)$ are the initial or starting values of $Z(t)$ and $Z(\tau)$ are the values of $Z(t)$ at an interval of time τ after $t = 0$ ($t(0)$). Now the initial conditions occur when $t = 0$ so

$$Z(0) = e^0 x(0) = x(0)$$

hence

$$Z(\tau) = x(0) + \int_0^\tau e^{-[A]t}[B]u\,dt$$

and substituting for $Z(\tau)$

$$e^{-[A]\tau}x(\tau) = x(0) + \int_0^\tau e^{-[A]t}[B]u\,dt$$

or

$$x(\tau) = e^{[A]\tau}x(0) + \int_0^\tau e^{[A](\tau-t)}[B]u\, dt \qquad (8.11)$$

We now have a method of calculating the entire state vector, that is, the current values of the process variables from a knowledge of the system relationships (vested in the matrix $[A]$) and the disturbing, and controlling inputs or stimuli $[B]u$.

The function of matrix $[A]$, $e^{[A]\tau}$, can be approximated by a series of powers of $[A]$

$$e^{[A]} = k_0[I] + k_1[A] + k_2[A]^2 + \cdots + k_{(n+1)}[A]^{(n+1)} \qquad (8.12)$$

It will be shown later that the characteristic equation of $[A]$ can be expressed in the form

$$\lambda^{(n)} + a_{(n-1)}\lambda^{(n-1)} + \cdots + a_1\lambda + a_0 = 0 \qquad (8.13)$$

where λ is any root of the characteristic equation. The Cayley-Hamilton theorem (proof of which is beyond the scope of this book) proves that the characteristic equation is always satisfied if the matrix $[A]$ is substituted for λ, thus

$$[A]^{(n)} + a_{(n-1)}[A]^{(n-1)} + \cdots + a_1[A] + a_0[I] = 0 \qquad (8.14)$$

Rearranging this equation

$$[A]^{(n)} = -a_{(n-1)}[A]^{(n-1)} \cdots a_1[A] - a_0[I] \qquad (8.15)$$

Multiplying through by $[A]$

$$[A]^{(n+1)} = -a_{(n-1)}[A]^{(n)} \cdots -a_1[A]^{(2)} - a_0[A] \qquad (8.16)$$

Substituting (8.14) in (8.16)

$$[A]^{(n+1)} = -a_{(n-1)}(-a_{(n-1)}[A]^{(n-1)} \cdots -a_1[A] - a_0[I]) \cdots$$
$$-a_1[A]^2 - a_0[A] \qquad (8.17)$$

Thus, it can be seen that all powers of $[A]$ greater than, or equal to n can be expressed as a linear sum of powers of $[A]^0, \ldots, [A]^{(n-1)}$, which means that the exponential function can be expanded as a *finite* series

$$e^{[A]t} = \alpha_0[I] + \alpha_1[A] + \alpha_2[A]^2 + \cdots + \alpha_{(n-1)}[A]^{(n-1)} \qquad (8.18)$$

where n is the order of the matrix $[A]$.

In any realizable system the matrix $[A]$ will *always* be square. Since the roots of the characteristic equation can be substituted for the matrix itself, there are n equations of the form

$$e^{\lambda t} = \alpha_0 + \alpha_1\lambda + \cdots + \alpha_{(n-1)}\lambda^{(n-1)}$$

which can be solved to find the values of the coefficients

$$\alpha_0, \alpha_1, \alpha_2, \ldots, \alpha_{(n-1)}$$

It will now be shown that these results enable an algebraic equation of the form

$$\mathbf{u} = [Y]\mathbf{x}$$

to be formed in which the matrix $[Y]$ is a model of the system in dynamic as well as steady-state terms.

If samples are taken at sufficiently frequent intervals to ensure that the disturbing inputs can be approximated to straight-line functions within any single interval, the integrations may be accomplished by taking the mean value of each input and the period as in Fig. 8.6. Over the period of the sampled interval, therefore, eqn. (8.11) becomes

$$\mathbf{x}_{(n+1)} = e^{[A]\tau}\mathbf{x}_{(n)} + [A]^{-1}(e^{[A]\tau} - I)[B]\mathbf{u}_{(n)} \qquad (8.19)$$

where τ is the sample interval time.

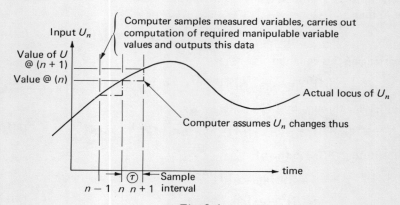

Fig. 8.6.

Assuming it to be possible, as in the last chapter, to drive the system to the desired state in one sample interval

$$\mathbf{x}_{(n+1)} = 0$$

where the states $x_1, x_2, x_3, \ldots, x_j$ are expressed as errors (desired value minus measured value) hence

$$\mathbf{x}_{(n)} = e^{-[A]\tau}[A]^{-1}(e^{[A]\tau} - [I])[B]\mathbf{u}_{(n)}$$

We can evaluate $e^{-[A]\tau}[A]^{-1}(e^{[A]\tau} - [I])$ to give a square matrix which we will call $[Y]$.

Then
$$\mathbf{x}_{(n)} = [Y][B]\mathbf{u}_{(n)} = [W]\mathbf{u}_{(n)} \qquad (8.20)$$

or
$$\mathbf{u}_{(n)} = [W]^{-1}\mathbf{x}_{(n)}$$

In general, for a n-dimensional system, there may not be as many as n manipulable inputs. In this case the equation can only be solved for as many states as there are manipulable inputs. The state equation may be 'partitioned' as shown*

$$\begin{bmatrix} x_1 \\ \vdots \\ x_k \\ \hline \vdots \\ x_n \end{bmatrix} = \begin{bmatrix} y_{11} & \cdots & y_{1n} \\ \vdots & & \vdots \\ y_{k1} & \cdots & y_{kn} \\ \vdots & & \vdots \\ y_{1n} & \cdots & y_{nn} \end{bmatrix} \times \begin{bmatrix} b_{11} & \cdots & b_{1k} & \cdots & b_{1n} \\ \vdots & & \vdots & & \vdots \\ b_{k1} & \cdots & b_{kk} & \cdots & b_{kn} \\ \vdots & & \vdots & & \vdots \\ b_{n1} & \cdots & b_{nk} & \cdots & b_{nn} \end{bmatrix} \times \begin{bmatrix} u_1 \\ \vdots \\ u_k \\ \hline \vdots \\ u_n \end{bmatrix} \qquad (8.21)$$

This yields the restricted equation of order k for k manipulable inputs

$$\begin{bmatrix} x_1 \\ \vdots \\ x_k \end{bmatrix} = -\begin{bmatrix} \tau_{11} & \cdots & \tau_{1h} & \cdots & \tau_{1k} \\ \vdots & & \vdots & & \vdots \\ \tau_{k1} & \cdots & \tau_{kh} & \cdots & \tau_{kk} \end{bmatrix} \begin{bmatrix} u_1 \\ \vdots \\ u_g \\ \vdots \\ u_k \end{bmatrix} \quad \text{or} \quad \mathbf{x} = -[R]\mathbf{u} \qquad (8.22)$$

where
$$\tau_{ij} = \sum_{\alpha=1}^{n} (b_{\alpha j} * y_{i\alpha})$$

hence
$$\mathbf{u}_{(n)} = -[R]^{-1}\mathbf{x}_{(n)} \qquad (8.23)$$

The partitioning of $[B]$ and $[Y]$ matrices leads to only part of the system response being considered. The remaining part of the system response is defined under two headings

1. The response to unconsidered (unmanipulable) inputs.
2. The response to all inputs in terms of the remaining system states.

The significance of 1 is that the control action is correctly classified as feed-

* The rows of $[Y]$ and columns of $[B]$ may have to be rearranged before partitioning.

back since only known errors are corrected. The significance of 2 is that some states are not controlled at all.

It should be noted that if there is only one manipulable input the state equation is reduced by the above partitioning to a scalar system. It should also be noted that the measurements of both inputs and system states are made relative to the steady-state 'desired values' of the states and the corresponding values of the inputs.

The steady-state values of the system states are the desired values, which are known, but what are the corresponding values of the inputs? Unless the closed-loop system is unstable it will converge to the steady-state desired vector; it follows therefore that over a period of time equivalent to many sample intervals the steady-state values of the driving inputs will be given by the integral of the measured values of the system states. Thus, the steady-state component of the driving inputs is given by the expression

$$\mathbf{u}_{ss} = -[R]^{-1} \textcircled{t} \int_{t_0}^{t} \mathbf{x}\, dt \qquad (8.24)$$

and the complete feedback algorithm in discrete form is

$$u_{(n)} = ([R]^{-1} x_{(n)} + \textcircled{t} [R]^{-1} \sum_{i=0}^{(n-1)} x_{(i)}) \qquad (8.25)$$

where $\textcircled{t}$ represents the integral time constant. The correspondence between this equation and the normal two term algorithm for a single dimension control system p. 78 is obvious. The third or derivative term of a conventional single parameter controller corresponds to varying the gain in proportion to the rate of change of the process variable; but the gain in this multivariable control algorithm has been defined as that which will result in the error vector being eliminated in one sample interval (assuming the model to be perfect). Hence a third term cannot readily be applied.

It will be recalled that feedforward action can be added by letting the measured value vector include measurable but non-manipulable inputs. This is achieved by replacing the measured value vector $\mathbf{x}$ in eqn. (8.25) above by

$$\mathbf{x} + \mathbf{x}' \quad \text{where } \mathbf{x}' = [S]\mathbf{u}'$$

and

$$\mathbf{u}' \begin{bmatrix} u_{(k+1)} \\ \vdots \\ u_p \end{bmatrix}$$

and

$$[S] = \begin{bmatrix} y_{11} & \cdots & y_{1n} \\ \vdots & & \vdots \\ y_{k1} & \cdots & y_{kn} \end{bmatrix} \begin{bmatrix} b_{1(k+1)} & \cdots & b_{1p} \\ \vdots & & \vdots \\ b_{n(k+1)} & \cdots & b_{np} \end{bmatrix}$$

Thus $\mathbf{x}'$ represents the estimated effect of the unmeasured inputs on the measured value vector.

9. Control of a real interactive process

This chapter is devoted to a worked example which is intended to illustrate how the theory presented in chapter 8 could be applied. No apology is made for using, once again, a refining column for this purpose. Refining forms a major part of any chemical complex and offers a splendid example of a multi-variable system; the continuity afforded allows comparisons to be made with the earlier steady state control example.

In a multistage refining process vapour and liquid pass up and down the column, in contradirection, passing through the liquid 'held' on the trays in consequence. Assuming that pressure is held constant, and making the assumption that specific and latent heats of the two components of the feed are equal, it is readily appreciated that the heat lost in condensing one mole of component A

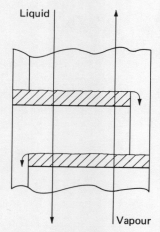

Fig. 9.4.

from the ascending vapour is approximately equal to that absorbed in vaporizing one mole of component B from the descending liquid. If the vapour is richer in component A than a vapour in equilibrium with the liquid on the tray, and the descending liquid poorer, then this is what will happen provided

120

that sufficient time and opportunity for contact between the liquid and vapour is allowed by the system design. This requirement dictates that vapour velocity is as low as practicable. The physical arrangement of a single stage shown in Fig. 9.4 is equivalent to two separate capacities or 'hold-ups' as shown in Fig. 9.5. The volume of component A can increase or decrease, provided that the total volume of A and B remains constant. Because of this constraint there is internal feedback from stage to stage in both the upward and the downward directions. The compositions on each tray will remain constant so long as the *ratio* of liquid flow down, and vapour flow up the column does not change. Any change in this ratio will start to affect the composition on the tray closest

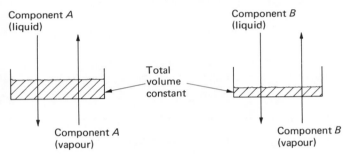

Fig. 9.5.

to the origin of the change first, the top tray if the liquid flowrate is disturbed, the bottom tray if the vapour flowrate is disturbed. Subsequently the disturbance will be propagated up or down the whole column subject to the seriesed time constant effects at each stage. The liquid flowrate is changed by a change in reflux ratio, the vapour by a change in 'boil-up' rate.

The feed, entering at some intermediate tray, splits the column into upper, or refining, section and lower or 'stripping' section. Any change in feed rate unbalances the vapour/liquid ratio in the lower section and, if some of the feed 'flashes', the upper section also. An increase or decrease in the boil-up rate will restore the balance in the lower section, but at the same time further disturb the balance in the upper section. The effect of both the original disturbance and the corrective action will propagate up the column, eventually affecting the distillate composition, which can, in turn, be corrected by a change in the reflux ratio, the results of which will eventually be felt at the feed tray, and later still at the bottom of the column. All these responses and counter responses are inevitably out of phase since they are 'transferred' through a different number of stages, out of sequence. Control action must be correct, in phase as well as amplitude, if the composition of either distillate or bottoms products is to remain constant in the face of feed disturbances. The effects of changes in feed composition are also propagated throughout the column, starting this time with the feed tray composition. Because the immediate effects of a feed flowrate change are transferred hydraulically, while those of a feed composition change are not, the latter are much 'slower' to propagate and, in consequence, more difficult to correct.

If both boil-up and reflux are manipulated in order to achieve the control objective, it is obvious that there is considerable danger of oscillatory response unless the two control actions are coordinated.

One way in which this may be achieved is to monitor the various effects as they propagate from tray to tray, and this can be done by measuring the temperature on each tray, as the temperature and *liquid* composition are dependent variables.

The equivalence of the transition matrix and the block diagram was discussed in chapter 8, and we will now use this technique to develop the model we require to implement this approach to system design. Considering liquid flow down the column first on its own gives the flow diagram Fig. 9.6 for a

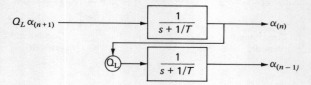

Fig. 9.6.

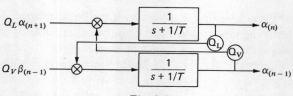

Fig. 9.7

single stage: superimposing the vapour flow gives Fig. 9.7. The defining equations are therefore

$$\dot{\alpha}_{(n)} = Q_L \alpha_{(n+1)} - \frac{1}{T} \alpha_{(n)} + Q_v \beta_{(n-1)}$$

and

$$\dot{\alpha}_{(n-1)} = Q_L \alpha_{(n)} - \frac{1}{T} \alpha_{(n-1)} + Q_v \beta_{(n-2)}$$

α = liquid composition
β = vapour composition

or in matrix form ($\delta = \beta/\alpha$)

$$\begin{bmatrix} \dot{\alpha}_{(n)} \\ \dot{\alpha}_{(n-1)} \end{bmatrix} = \begin{bmatrix} Q_L + \dfrac{1}{T} + \delta_{(n-1)}Q_v + & 0 \\ 0 + Q_L + & \dfrac{1}{T} + \delta_{(n-2)}Q_v \end{bmatrix} \begin{bmatrix} \alpha_{(n+1)} \\ \alpha_{(n)} \\ \alpha_{(n-1)} \\ \alpha_{(n-2)} \end{bmatrix} \qquad (9.11)$$

The matrix is obviously incomplete and if extended to include higher and lower

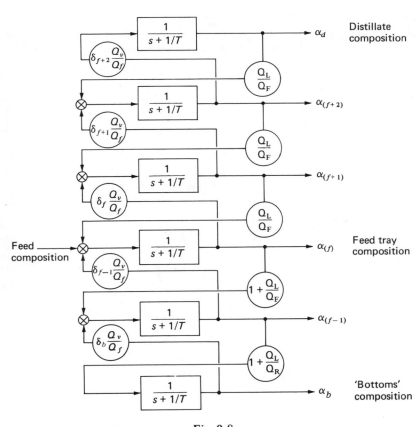

Fig. 9.8.

trays will have the general form above and below the feed tray 'n'.

$$\begin{bmatrix} \dot{\alpha}_{(n+2)} \\ \dot{\alpha}_{(n+1)} \\ \dot{\alpha}_{(n)} \\ \dot{\alpha}_{(n-1)} \\ \dot{\alpha}_{(n-2)} \end{bmatrix} = \begin{bmatrix} \frac{1}{T}+\delta_{(n+1)}Q_v & + & 0 & + & 0 & + & 0 \\ Q_L & + & \frac{1}{T}+\delta_{(n)}Q_v & + & 0 & + & 0 \\ 0 & + & Q_L & + & \frac{1}{T}+\delta_{(n-1)}Q_v & + & 0 \\ 0 & + & 0 & + & (Q_L+Q_F) & + & \frac{1}{T}+\delta_{(n-2)}Q_v \\ 0 & + & 0 & + & 0 & (Q_L+Q_F) & + & \frac{1}{T} \end{bmatrix} \begin{bmatrix} \alpha_{(n+2)} \\ \alpha_{(n+1)} \\ \alpha_{(n)} \\ \alpha_{(n-1)} \\ \alpha_{(n-2)} \end{bmatrix}$$

(9.12)

Adding the distillate and bottoms compositions into this equation and substituting Q_v/Q_F and Q_L/Q_F for Q_v and Q_L respectively (assuming hydraulic effects to be propagated instantly) the total free response is obtained (Fig. 9.8). Hence:

$$\begin{bmatrix} \dot{\alpha}_{(d)} \\ \dot{\alpha}_{(f+2)} \\ \dot{\alpha}_{(f+1)} \\ \dot{\alpha}_{(f)} \\ \dot{\alpha}_{(f-1)} \\ \dot{\alpha}_{(b)} \end{bmatrix} = \begin{bmatrix} \frac{1}{T}+\delta_{(d)}\frac{Q_v}{Q_F} & + & 0 & + & 0 & + & 0 & + & 0 \\ \frac{Q_L}{Q_F} & + & \frac{1}{T}+\delta_{(f+1)}\frac{Q_v}{Q_F} & + & 0 & + & 0 & + & 0 \\ 0 & + & \frac{Q_L}{Q_F} & + & \frac{1}{T}+\delta_{(f+1)}\frac{Q_v}{Q_F} & + & 0 & + & 0 \\ 0 & + & 0 & + & \frac{Q_L}{Q_F} & + & \frac{1}{T}+\delta_{(f)}\frac{Q_v}{Q_F} & + & 0 \\ 0 & + & 0 & + & 0 & + & \left(\frac{Q_L}{Q_F}+1\right) & + & \frac{1}{T}+\delta_{(f-1)}\frac{Q_v}{Q_F} \\ 0 & + & 0 & + & 0 & + & 0 & + & \left(\frac{Q_L}{Q_F}+1\right) & + & \frac{1}{T} \end{bmatrix} \begin{bmatrix} \alpha_{(d)} \\ \alpha_{(f+2)} \\ \alpha_{(f+1)} \\ \alpha_{(f)} \\ \alpha_{(f-1)} \\ \alpha_{(b)} \end{bmatrix}$$

(9.13)

An approximate solution to the state space equation can be obtained as follows.

Let the sample interval be T, then the mean rate of change over the interval will be

$$\frac{\Delta\alpha_{(n)}}{T} \quad \text{where } \Delta\alpha_{(n)} = \alpha_{DV} - \alpha_{(n)} \quad (\alpha_{DV} \text{ is the desired value})$$

where $\Delta\alpha_{(n+1)}$ and $\Delta\alpha_{(n-1)}$ represent changes of composition of the liquid held on different trays over one interval. Then the equation for tray n can be written

$$\frac{\Delta\alpha_{(n)}}{\tau} = \left(\frac{Q_L^*}{Q_F} + \Delta\frac{Q_L}{Q_F}\right)(\alpha_{(n+1)} + \tfrac{1}{2}\Delta\alpha_{(n+1)}) - \frac{1}{T}(\alpha_{(n)} + \tfrac{1}{2}\Delta\alpha_{(n)})$$

$$+ \delta_{(n-1)}\left(\frac{Q_v^*}{Q_F} + \Delta\frac{Q_v}{Q_F}\right)(\alpha_{(n-1)} + \tfrac{1}{2}\Delta\alpha_{(n-1)}) \quad (9.14)$$

where $\Delta(Q_L/Q_F)$ and $\Delta(Q_V/Q_F)$ are increments in the inputs to the system with respect to the steady-state integrated values Q_L^*/Q_F and Q_V^*/Q_F which will eliminate $\Delta\alpha_{(n)}$ in exactly one interval. The estimated value of these increments is to be calculated.

At the end of the process sample interval the value of $\alpha_{(n)}$ will be α_{DV}, hence the middle term of (9.14) can be re-written

$$-\frac{1}{T}(\alpha_{DV} - \tfrac{1}{2}\Delta\alpha_{(n)})$$

Equation (9.14) can now be rearranged

$$\left(\frac{1}{\tau} - \frac{1}{2T}\right)\Delta\alpha_{(n)} + \frac{1}{T}\alpha_{DV} =$$

$$\left(\frac{Q_L^*}{Q_F} + \Delta\frac{Q_L}{Q_F}\right)(\alpha_{(n+1)} + \tfrac{1}{2}\Delta\alpha_{(n+1)}) + \delta_{(n-1)}\left(\frac{Q_V^*}{Q_F} + \Delta\frac{Q_V}{Q_F}\right)\alpha_{(n-1)} + \tfrac{1}{2}\Delta\alpha_{(n-1)})$$

(9.15)

Now if the measured values are measured relative to the steady-state desired values and the inputs relative to the integrated steady-state values as on p. 118. then

$$\alpha_{DV} = 0, \quad \Delta\alpha_{(n)} = \alpha_{(n)} \quad \text{and} \quad \Delta\frac{Q_L}{Q_F} = \frac{Q_L}{Q_F},$$

$$\Delta\frac{Q_V}{Q_F} = \frac{Q_V}{Q_F}, \quad \text{and} \quad \frac{Q_L^*}{Q_F} = \frac{Q_V^*}{Q_F} = 0 \quad (9.16)$$

and (9.15) reduces to

$$\left(\frac{1}{\tau} - \frac{1}{2T}\right)\alpha_{(n)} = \frac{Q_L}{Q_F}(\alpha_{(n+1)} + \tfrac{1}{2}\alpha_{(n+1)}) + \delta_{(n+1)}\frac{Q_L}{Q_F}(\alpha_{(n-1)} + \tfrac{1}{2}\alpha_{(n-1)})$$

If τ is much smaller than T

$$\left(\frac{1}{\tau} - \frac{1}{2T}\right) \simeq \frac{1}{\tau}$$

125

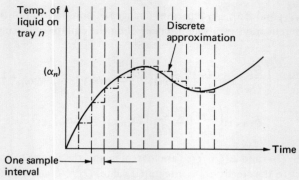

Fig. 9.9.

and (9.16) reduces to *

$$\frac{\alpha_{(n)}}{(T)} = \frac{Q_L}{Q_F}(\alpha_{(n+1)} + \tfrac{1}{2}\Delta\alpha_{(n+1)}) + \delta_{(n-1)}\frac{Q_V}{Q_L}(\alpha_{(n-1)} + \tfrac{1}{2}\Delta\alpha_{(n-1)})$$

Now Q_L/Q_F and Q_V/Q_F, being computed variables, can be as large as necessary within the limits set by control valve size, etc., but the compositions on the trays can alter very little over a single process sample interval, which is much shorter than the tray time constant. Thus to a first approximation $(\alpha_{(n+1)} + \tfrac{1}{2}\Delta\alpha_{(n+1)}) = \alpha_{(n+1)}$ with little loss of accuracy (note: these are not the measured values of the system). Hence (9.16) becomes

$$\frac{\alpha_{(n)}}{(T)} = \frac{Q_L}{Q_F}\alpha_{(n+1)} + \delta_{(n-1)}\frac{Q_V}{Q_F}\alpha_{(n-1)}$$

or

$$\alpha_{(n)} = k_1\alpha_{(n+1)}\frac{Q_L}{Q_F} + k_2\alpha_{(n-1)}\frac{Q_V}{Q_F} \text{ for tray } n$$

where k_1 and k_2 are constants. Similarly

$$\alpha_{(m)} = k_3\alpha_{(m+1)}\frac{Q_L}{Q_F} + k_{(4)}\alpha_{(m-1)}\frac{Q_V}{Q_F} \text{ for tray } m.$$

The restricted form of the equation can now be written

$$\begin{bmatrix}\alpha_{(n)} \\ \alpha_{(m)}\end{bmatrix}_{(i)} = \begin{bmatrix}k_1\alpha_{(n+1)} & k_2\alpha_{(n-1)} \\ k_3\alpha_{(m+1)} & k_4\alpha_{(m-1)}\end{bmatrix}\begin{bmatrix}\dfrac{Q_L}{Q_F} \\ \dfrac{Q_V}{Q_F}\end{bmatrix}_{(i)}$$

or

$$\mathbf{x}_{(i)} = R\mathbf{u}_{(i)}$$

for a system defined in terms of two measured variables—two tray compositions—and two manipulated variables—reflux/feed flowrate, and boil-up/feed flowrate. The constants k_1 and k_3 include $\delta_{(n)}$ and $\delta_{(m)}$. For a large system (of, say, 40 trays) the elements of the matrix can now be approximated to constants for any given set of steady-state temperatures, since the temperatures of the trays immediately above and below the two control trays can only alter slowly due to the hold-up on each tray.

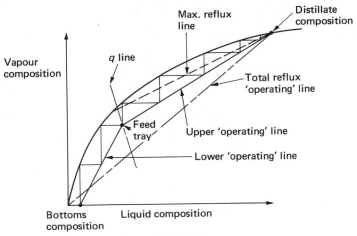

Fig. 9.10.

The *modus operandi* of the scheme is best appreciated with reference to the McCabe–Thiele diagram Fig. 9.10. This shows that if the top tray and the feed tray are selected, the upper operating line is 'fixed' since, under the assumption leading to this diagram, these lines are straight. The lower operating line on the other hand is free to move, and must do so in order to satisfy the composition balance. Response to any disturbance is thus restricted to the 'stripping' section in terms of composition.

The assumptions made in the course of the above development should be fully understood. Approximation of tray temperatures to discrete functions is valid only if the sample period is short relative to the rate at which these change: on the other hand the assumption that hydraulic responses take place instantaneously is valid only if the sample interval is sufficiently long. These conflicting conditions can normally be easily met in practice. The assumptions implicit in the McCabe–Thiele diagram are not fully reflected in practice, the operating lines are never straight. The slope, however, cannot change sign, since this would imply that at least one tray was operating at a temperature lower than the one below it.

10. System identification

Unfil now it has been assumed that the system, simple or complex, *can* be defined in mathematical terms by considering the physical laws to which it is subject, and the dimensions of the system. This is, of course, not always possible because the physical laws are not sufficiently well understood or because such a task is simply too difficult. We require a means of discovering the system definition by testing the actual system or by testing existing systems in order to establish modifications required in a new system to be built.

Any means of testing a system must basically consist of stimulating it and discovering the relationship of the response to the stimulus. Few process systems can be operated purely for test purposes and, in any case, if artificial stimulus alters the normal operation, the behaviour of the system described by the stimulus/response relationship obtained may not be typical of the system. Sometimes normal operations provide suitable perturbations; alternatively we must evolve a method of superimposing a stimulus on the normal stimuli of the functioning system in such a way that the normal behaviour is not significantly modified.

10.1 Identification of multivariable systems

In previous chapters the model of the system has been defined as the matrix relating the required input vector whose elements are the errors in the system states. It has been shown that this matrix, or a suitable approximation to it, can be synthesized from knowledge of the physical relationships of the system. If it is possible to learn from the behaviour of the system when it is disturbed, it is possible to update whatever initial estimate is available for a linear (constant matrices) system. While this subject is really beyond the scope of this book, the principles involved will be discussed in this chapter and a simulated example shown, in order to illustrate one of the strongest incentives to the adoption of the digital computer for real-time process control. The ability of even a small computer to analyse information in a short time makes it possible to implement on line identification techniques in real time. As a direct result, a self-optimizing control system becomes a practical possibility.

In implementing multivariable control, we are controlling the process itself, and not just the environment in which it takes place. Few chemical processes are today automatically controlled. Such controls as are used are for the purpose of stabilizing individual variables (level, pressure, etc.) which are the 'dimensions' of the system or process. Control of the process itself is vested in the human operator and the model of the process (which, as we have seen, is necessary for control) is also vested in him. He learns by experience to anticipate the response to any adjustment he may make to the 'dimensions' of the process: in doing so he implements both feedforward and feedback control, taking into consideration the interactions of the system. Such knowledge is, of course, imprecise and variable; he can, at best, make a subjective and very slow analysis of the plant behaviour by correlating the information he reads on his instrument displays; his memory is also inaccurate. A computer, if correctly programmed, can make a very rapid, accurate, and complete analysis, and can remember and apply this with equal accuracy.

In order to learn from the behaviour of the system, it must be made to respond in a *measurable* fashion to *known* inputs, (disturbances) (a perfectly controlled system can provide no information since the object of control is to suppress response to unwanted but unavoidable disturbances). If the process is stable it will have *to* be stimulated in order to obtain information. If the system is less stable it will obviously require less stimulation.

If we measure a three-dimensional object, such as a cardboard carton, with a single-dimensional scale, a ruler, it is necessary to obtain three measurements in order to fully 'define' the carton. The measurements we take do not have to be at right-angles to each other necessarily, although this is ideal; however, they must be taken in directions which are sufficiently different that they 'span' the space occupied by the carton. Similarly, the measurements, and therefore the disturbances which result in the responses measured, must 'span' the 'system space'. If an independent (non-interacting) dimension of a system is disturbed (such as the water level in a boiler drum) we cannot expect the response to tell us anything about the system's response to disturbances in other dimensions (steam condition, for instance). Even if the disturbance comprises a component of each of the systems 'dimensions' the response is only defined for that particular 'combination'. In order to identify an n-dimensional system, it is necessary to apply n separate disturbances (in sequence) to 'span' the system state space. It may be possible under certain conditions to make use of disturbances to which the systems is necessarily subjected, but as these are unpredictable, this is unlikely. It is unlikely, moreover, that these will 'span' the system space fully.

In chapter 8, p. 117, it was shown that the discrete form of the state equation can be written

$$\mathbf{x}_{(n+1)} = [F]\mathbf{x}_{(n)} - [G]\mathbf{u}_{(n)}$$

where

$$[F] = e^{[A]T} \qquad (10.1)$$

and
$$G = [A]^{-1}(e^{-[A]T} - [I])[B]$$

In the absence of any unknown inputs, an output vector **u** can be computed such that $\mathbf{x}_{(n+1)} = \mathbf{0}$. Substituting this in (10.1) gives

$$\mathbf{x}_{(n)} = -[Y]\mathbf{u}_{(n)} \tag{10.2}$$

where

$$[Y] = e^{-[A]T}[A]^{-1}(e^{-[A]T} - [T])[B]$$

Defining some estimate of $[Y]$ say $[E]$, a matrix of the same order, the corresponding equation

$$\mathbf{x}_{(n)} = -[E]\mathbf{u}'_{(n)} \tag{10.3}$$

can be solved to give an estimate of the vector **u**, **u**'. Combining (10.2) and (10.3):

$$\mathbf{x} - \mathbf{x}' = ([Y] - [E])\mathbf{u}' \tag{10.4}$$

Defining an error matrix $[\phi] = ([Y] - [E])$ the elements of which are the difference between an element of $[Y]$ and the corresponding element of $[E]$

$$\mathbf{x} - \mathbf{x}' = [\phi]\mathbf{u}' \tag{10.5}$$

or

$$\mathbf{x}' = [\phi]\mathbf{u}' + \mathbf{x}$$

or

$$\mathbf{x}' = -[E]\mathbf{u}' - [\phi]\mathbf{u}'$$

hence

$$\mathbf{x} = -[E]\mathbf{u} - [\phi]\mathbf{u} \tag{10.6}$$

The ideal system is therefore defined as the sum of the estimate 'system' and the error 'system'. The term $(\mathbf{x} - \mathbf{x}')$ is equivalent to the vector difference between the intended (estimated) change in the state vector and the actual change, the 'error vector'. The n **x** vectors, and n **u** vectors obtained after n intervals form two n-dimensioned matrices from which two equations can be formed to solve for the two *rows* of $[\phi]$. The procedure can best be seen from an example. It will be assumed that, despite, self-regulation, the roots of the characteristic equation are complex, though not positive real (unstable). At the beginning of the first interval the desired value vector is set to a suitable value with respect to the 'present' value of the measured value vector; at the beginning of the second interval it is reset to a value 'orthogonal' to the first.

The state equation is

$$\begin{vmatrix} \dot{x}_1 \\ \dot{x}_2 \end{vmatrix} = \begin{vmatrix} -\dfrac{1}{T_1} & +a \\ +b & -\dfrac{1}{T_2} \end{vmatrix} \begin{vmatrix} x_1 \\ x_2 \end{vmatrix} + \begin{vmatrix} 1 & 0 \\ 0 & 1 \end{vmatrix} \begin{vmatrix} u_1 \\ u_2 \end{vmatrix} \qquad (10.7)$$

Let T_1 and T_2 be 1, let a and $b = 1$, then the characteristic equation is

$$\left. \begin{array}{c} s^2 + 2s + 2 = 0 \\ [s+(1+j)][s+(1-j)] = 0 \end{array} \right\} \equiv \begin{bmatrix} -2 & -1 \\ +2 & 0 \end{bmatrix} \qquad (10.8)$$

or

Thus, as shown in chapter 10, the equation

$$e^{\lambda *} = \alpha_0 + \alpha_1 \lambda$$

must be satisfied for both roots. Selecting a sample interval $\widehat{T} = 0.1$ minute

$$e^{0.1} + e^{0.1j} = \alpha_0 + (\alpha_1 + \alpha_1 j) \qquad (10.9)$$

and

$$e^{0.1} + e^{-0.1j} = \alpha_0 + (\alpha_1 - \alpha_1 j) \qquad (10.10)$$

Hence

$$\alpha_1 = \sin(0.1 \text{ rad}) = \sin(5.729°) = 0.0998$$

and

$$\alpha_0 = e^{-0.1} + \cos(5.729°) - (0.0998) = 1.8000$$

Hence $[F] = e^{[A]}$ is given by

$$\begin{bmatrix} 1.8000 & 0 \\ 0 & 1.8000 \end{bmatrix} + \begin{bmatrix} -0.1996 & -0.0996 \\ 0.1996 & 0 \end{bmatrix} = \begin{bmatrix} 1.6004 & -0.0998 \\ 0.1996 & +1.8000 \end{bmatrix} \qquad (10.11)$$

and $[G] = [A]^{-1}([F] - [I])$ by

$$\tfrac{1}{2} \begin{bmatrix} 0 & 1 \\ -2 & -2 \end{bmatrix} \begin{bmatrix} 0.6004 & -0.0998 \\ -0.1996 & +0.8000 \end{bmatrix} = \begin{bmatrix} 0.0998 & +0.4000 \\ -0.8000 & -0.7002 \end{bmatrix} \qquad (10.12)$$

and $[Y]$ is

$$0.3447 \times \begin{bmatrix} 1.8000 & +0.0998 \\ -0.1996 & +1.6004 \end{bmatrix} \begin{bmatrix} 0.0998 & +0.4000 \\ -0.8000 & -0.7002 \end{bmatrix} = \begin{bmatrix} 0.0344 & +0.2241 \\ -0.4482 & -0.4137 \end{bmatrix} \qquad (10.13)$$

We require an initial estimate of the matrix $[Y]$ which is defined as $[E]$, and in order to illustrate the method we will choose this to be as different from the 'real system as is reasonable'.

Let the estimate system be

$$s^2 + 5s + 6 = 0$$

or

$$(s + 2)(s + 3) = 0$$

For this system the roots are real and negative, the damping coefficient being 1.02. The matrix $[E]$ is obtained in exactly the same manner as was $[Y]$ above.

$$[E] = \begin{vmatrix} 0.0773 + 0.0126 \\ -0.0084 + 0.0984 \end{vmatrix} \quad (10.14)$$

The behaviour of the 'real' system can be calculated from $[F]$ and $[G]$ for inputs computed using $[E]$. In this way the error at the end of any interval is the value of the state vector calculated from $[F]$ and $[G]$. First interval

$$[E]^{-1} = \begin{vmatrix} 10.023 + 1.633 \\ -1.089 + 12.759 \end{vmatrix}$$

where $\begin{vmatrix} 0.01 \\ 0.01 \end{vmatrix}$ is the desired value vector, hence

$$\begin{vmatrix} Q_B \\ Q_W \end{vmatrix}_{(1)} = - \begin{vmatrix} 10.023 + 1.633 \\ -1.089 + 12.759 \end{vmatrix} \begin{vmatrix} 0.01 \\ 0.01 \end{vmatrix} = \begin{vmatrix} -1.1656 \\ -1.1670 \end{vmatrix}$$

Final state

$$\begin{vmatrix} x_1 \\ x_2 \end{vmatrix}_{(1)} = \begin{vmatrix} 1.6004 - 0.0998 \\ 0.1996 + 1.8000 \end{vmatrix} \begin{vmatrix} 0.01 \\ 0.01 \end{vmatrix} + \begin{vmatrix} 0.0998 + 0.4000 \\ -0.8000 - 0.7002 \end{vmatrix} \begin{vmatrix} -1.1656 \\ -1.1670 \end{vmatrix}$$

$$= \begin{vmatrix} 0.1501 \\ 0.1996 \end{vmatrix} + \begin{vmatrix} -0.5831 \\ 1.7496 \end{vmatrix}$$

$$\begin{vmatrix} -0.4330 \\ +1.9492 \end{vmatrix}$$

Second interval

where $\begin{vmatrix} 0.01 \\ -0.01 \end{vmatrix}$ is the desired value vector, hence

$$\begin{vmatrix} u_1 \\ u_2 \end{vmatrix}_{(2)} = \begin{vmatrix} 10.023 + 1.633 \\ -1.089 + 12.759 \end{vmatrix} \begin{vmatrix} 0.01 \\ -0.01 \end{vmatrix} = \begin{vmatrix} -0.8390 \\ +1.3848 \end{vmatrix}$$

Final state

$$\begin{vmatrix} \dot{x}_1 \\ \dot{x}_2 \end{vmatrix}_{(2)} = \begin{vmatrix} 1.6004 & -0.0998 \\ 0.1996 & +1.8000 \end{vmatrix} \begin{vmatrix} 0.01 \\ -0.01 \end{vmatrix} + \begin{vmatrix} 0.0998 & +0.400 \\ -0.8000 & -0.7002 \end{vmatrix} \begin{vmatrix} -0.8390 \\ +1.3848 \end{vmatrix}$$

$$= \begin{vmatrix} 0.1700 \\ -0.1600 \end{vmatrix} + \begin{vmatrix} 0.4702 \\ -0.2984 \end{vmatrix}$$

$$= \begin{vmatrix} 0.6202 \\ -0.4584 \end{vmatrix}$$

Identification calculation

$$[U] = \begin{vmatrix} -1.1656 & -1.1670 \\ -0.8390 & +1.3848 \end{vmatrix} \quad \text{hence} \quad [U]^{-1} = \begin{vmatrix} -0.5340 & -0.4500 \\ -0.3235 & +0.4494 \end{vmatrix}$$

and

$$\begin{vmatrix} m_{11} \\ m_{12} \end{vmatrix} = \begin{vmatrix} -0.5340 & -0.4500 \\ -0.3235 & +0.4494 \end{vmatrix} \begin{vmatrix} -0.4330 \\ +0.6202 \end{vmatrix} = \begin{vmatrix} -0.0479 \\ +0.4188 \end{vmatrix}$$

also

$$\begin{vmatrix} m_{21} \\ m_{22} \end{vmatrix} = \begin{vmatrix} -0.5340 & -0.4500 \\ -0.3235 & +0.4494 \end{vmatrix} \begin{vmatrix} -0.8346 \\ -0.8366 \end{vmatrix} = \begin{vmatrix} -0.8346 \\ -0.8366 \end{vmatrix}$$

Thus

$$[Y] = [E] + [\phi] = \begin{vmatrix} 0.0773 & +0.0126 \\ -0.0084 & +0.0984 \end{vmatrix} + \begin{vmatrix} -0.0479 & +0.4188 \\ -0.8346 & -0.8366 \end{vmatrix}$$

$$= \begin{vmatrix} 0.0294 & +0.4314 \\ -0.8430 & -0.7381 \end{vmatrix} \tag{10.15}$$

The scale of this matrix is clearly inaccurate; this is because the calculation is highly dependent on the values of the determinants of the various matrices computed. It is found that this scalar inaccuracy increases as the damping factor of the system decreases. However the correct gain or scale can be found by trial and error methods after identification in exactly the same way as it normally is for a single variable control loop. In this case the scale must be reduced by a factor of approximately 2, giving

$$\begin{vmatrix} 0.0147 & +0.2157 \\ -0.4215 & -0.3691 \end{vmatrix} \tag{10.16}$$

which is very close to the 'real value of $[Y]$. If the final values of the state vector at the end of the two intervals is recalculated, it is seen that the errors under the new 'model' are very small indeed.

$$\begin{vmatrix} x_1 \\ x_2 \end{vmatrix}_{(1)} = \begin{vmatrix} 0.0143 \\ 0.0096 \end{vmatrix}$$

$$\begin{vmatrix} x_1 \\ x_2 \end{vmatrix}_{(2)} = \begin{vmatrix} -0.0024 \\ +0.0296 \end{vmatrix}$$

It is important to remember that the **x** and **u** vectors are measured with respect to the appropriate steady-state values; since the latter will change when the desired value vector is changed, each of the *n* tests must commence from steady-state. It is obviously no easy matter to identify a system of many dimensions.

10.2 Other methods

While the identification of a stable, well-damped multivariable process can be achieved by the method outlined, it is unlikely that a highly oscillatory process would tolerate the imposition of artificial disturbances in practice. Other methods of identification have been developed for single input/output systems which avoid this disadvantage, the use of which will undoubtedly be extended to multivariable systems.

Provided that the system itself is not changing with the passage of time *any* stimulus will produce a unique response. We know that *any* system will modify the stimulus in some fashion that involves changes of magnitude and of phase. The most obvious stimuli to apply is a sinusoid, since, as discussed on p. 30, the response must also be sinusoidal, after initial transients have decayed, and the relationship of response to stimulus can be expressed as a single complex number. We have already seen that the transfer function of the system can be synthesized from a series of such tests, each test discovering the response to a single frequency of stimulus. However, such a technique has obvious limitations; the necessary series of tests is time consuming and the normal operation of the process is certain to be affected, often seriously, by stimuli of this nature.

Various techniques have been evolved to test systems with a single pulse of suitable shape based on the fact that any pulse can be decomposed into constituent frequencies of different amplitudes. Such a pulse must comprise, in sufficient amplitude, all the frequencies to which a significant response can be obtained from the system under tests. It is possible to decompose, by Fourier analysis, both stimulus and response, and to obtain the relationship for each frequency as a complex quantity thus achieving in one test all that frequency response achieves in a series of tests. While these techniques are a great advance on frequency testing the fact remains that the normal behaviour of very many process systems, particularly multivariable non-independent systems, is seriously disturbed by the superimposition of a pulse of such duration and amplitude as to permit the analysis.

It has been shown that the ratio of response to stimuli expressed in integral transform terms describes the behaviour of a system, that is, describes the relationship of the system output to the input which causes it, but there are other ways of expressing the relationship. The two functions plotted on the *same* time scale might be as shown in Fig. 10.1.

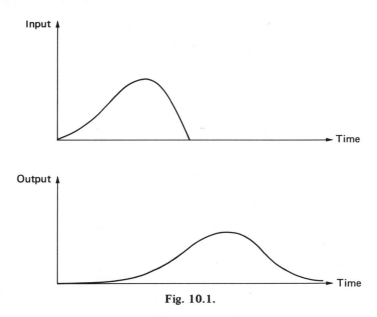

Fig. 10.1.

If now the time scale of one function is displaced from the time scale of the other by an interval as shown in Fig. 10.2, we can take the product of the input function and the output function, for each instant in time as τ is varied from $+\infty$ to $-\infty$ and obtain a *unique* function of τ. The function is unique in that it is the *only* function which can result from this procedure; the same function *could*, however, be obtained with other input and output functions. This function is known as the 'correlation function' of the two functions we began with, and, since for any two given functions only one 'correlation function' is possible, it is a sufficient description of the system's behaviour.

The correlation function of two functions of time $x(t)$ and $y(t)$ is defined as the average overall time of the product of the two functions as their relative position in time τ changes

$$\phi(x, y)_{(\tau)} = \underset{T \to \infty}{\text{Limit}} \frac{1}{2T} \int_{-T}^{\mp T} x(t) y(t + \tau) \, dt \qquad (10.17)$$

where $\phi(x, y)_{(\tau)}$ is the 'cross' correlation function of time functions $x(t)$ and

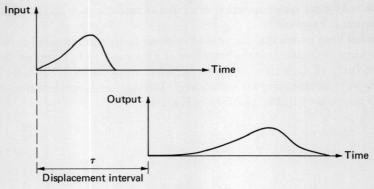

Fig. 10.2.

$y(t)$. τ is the measure of relative position in time. In practice τ can be achieved by imposing variable delays on the input and output functions to give $-\tau$ or $+\tau$.

To illustrate the concept of correlation functions consider a square wave function correlated with itself—the autocorrelation function—as shown in Fig. 10.3.

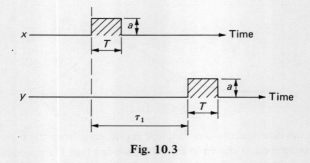

Fig. 10.3

1. At interval or delay τ_1 the product xy is zero at all points in time because when x has a finite value a, y is zero and when $y = a$, $x = 0$ (Fig. 10.4).

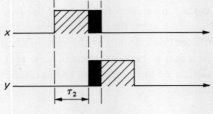

Fig. 10.4

2. At delay τ_2 the value of the correlation function $\phi(x, y)_{(\tau)}$ is the product of the shaded area.

3. At delay $\tau_3 = 0$ the correlation function is a maximum.
4. As τ becomes negative (x delayed with respect to y) the correlation function decreases.

Thus the autocorrelation function of the square wave function of amplitude a and duration T is a shown in Fig. 10.5.

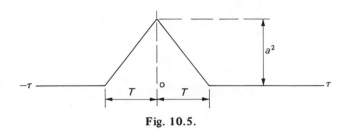

Fig. 10.5.

If we imagine that the same pulse is applied to stimulate a given system and that the resultant response is a square wave pulse attenuated to $\tfrac{1}{2}a$ amplitude and delayed τ units of time, but otherwise unmodified, the *cross*-correlation of the input and output functions is as shown in Fig. 10.6.

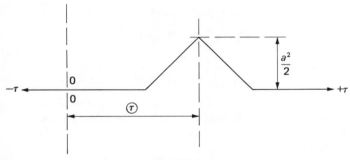

Fig. 10.6

Since the correlation function of two time functions, stimulus and response, is unique and the transfer function is also unique there must be a unique relationship between the two. If the stimulating function is truly random comprising, by definition all frequencies $+\infty$ to $-\infty$ Hz in equal amplitudes it can be shown that the Laplace transform of the cross-correlation function *is* the transfer function of the system.

By definition the CCF of stimulus $x(t)$ and response $y(t)$ is

$$\phi(xy)_{(\tau)} = \underset{T \to \infty}{\text{Limit}} \frac{1}{2T} \int_{-T}^{+T} f(x)_{(t)} f(y)_{(t+\tau)} \, dt$$

The frequency spectrum of

$$f(y)(t+\tau) \text{ is } \sum \bigcirc\!\!\!\!y \; n(j\omega) \, e^{nj\omega(t+\tau)}$$

Substituting this in the expression for $\phi(x,y)_{(\tau)}$ we get

$$\phi(x,y) = \underset{T\to\infty}{\text{Limit}} \frac{1}{2T} \int_{-T}^{+T} \sum \bigcirc\!\!\!\!y \; n(j\omega) \, e^{nj\omega(t+\tau)} f(x)(t) \, dt \quad (10.18)$$

$$= \underset{T\to\infty}{\text{Limit}} \sum \bigcirc\!\!\!\!y \; n(j\omega) \, e^{nj\omega\tau} \frac{1}{2T} \int_{-T}^{+T} f(x)(t) \, e^{n(j\omega t)} \, dt$$

But

$$\frac{1}{2T} \int_{-T}^{+T} f(x)(t) \, e^{n(j\omega)t} \, dt = \bigcirc\!\!\!\!x \; n(-j\omega)$$

Hence

$$\phi(x,y)(\tau) = \sum \bigcirc\!\!\!\!y \; n(j\omega) \bigcirc\!\!\!\!x \; n(-j\omega) \, e^{n(j\omega)\tau} \quad (10.19)$$

By taking limits in the same way as on p. 32, we can show that

$$\phi(x,y)_{(\tau)} = \frac{1}{2\pi j} \int_{-j\omega}^{+j\omega} F(y)_{(j\omega)} F(x)_{(-j\omega)} \, e^{(j\omega)\tau} \, d(j\omega) \quad (10.20)$$

Hence $\Phi(x,y)_{(j\omega)}$, the Fourier transform of

$$\phi(x,y)_{(j\omega)} = \int_{-\infty}^{\infty} \phi(x,y)_{(\tau)} \, e^{(-j\omega)\tau} \, d\tau$$

and

$$\Phi(x,y)_{(-j\omega)} = F(y)_{(j\omega)} F(x)_{(-j\omega)} \quad (10.21)$$

Now $F(x)_{(-j\omega)}$ the Fourier transform of the input must be unity for a truly random function since a truly random function can be defined as one comprising all frequencies in equal amplitude.

Hence for a truly random stimulus

$$\Phi(x,y)_{(j\omega)} = F(y)_{(j\omega)} \quad (10.22)$$

which, since the input function is unity, is the Fourier transform version of the transfer function. Replacing $(j\omega)$ by s the Laplace operator yields the transfer

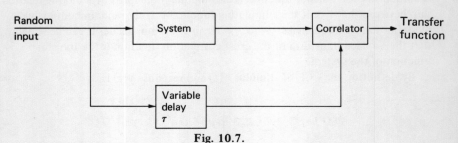

Fig. 10.7.

function, and the Laplace transform of the CCF *is* the system transfer function.

Thus if a system is forced or stimulated by a truly random input and the response correlated with the input for varying delays, as shown in Fig. 10.7, we obtain the transfer function of the system. Since random perturbations are equivalent to 'noise' which any system must be capable of accepting within reasonable amplitude limits the superimposition of such a disturbance for test purposes on an operating plant is unlikely to affect product quality, yield or throughput significantly.

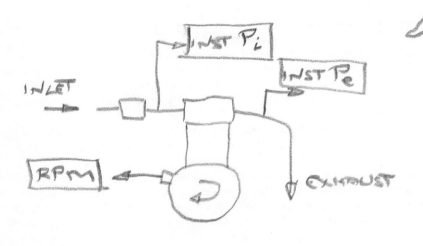

CROSS - CORRELATE RPM WITH P_i + P_e TO DETERMINE THE EFFECT OF CHANGES (P_i ALONE WOULD DO)

11. On line computer control today

During the later chapters of this book considerable emphasis is laid on the use of the digital computer for real time on line control of the process itself. Indeed, the techniques described for control of interactive systems are not otherwise possible. It is therefore apposite to discuss briefly the place of the computer in process control today.

There is a tendency for management to view the introduction of computer control techniques with a degree of scepticism which, while it cannot be justified logically, is nevertheless very understandable. Management, even if technically qualified is seldom able to make a realistic judgement of the real economic benefits to be expected. The techniques currently being developed in universities and research establishments are often far ahead of the ability of practising engineers to translate into working, maintainable plant. The development of the digital computer during the 'fifties and early 'sixties was aimed at business applications: in the main the requirements were for extensive easily accessible data systems, the computational ability of the computer being of relatively minor importance. Attempts made during these years to harness such machines to real time on-line control were usually disastrous, and when it is realized that it was during this same period that almost the whole of modern control theory developed, this is not really surprising. Management is right to be very cautious in the light of such a history and the computer manufacturers should not now complain if business is hard to find.

The position however has recently changed fundamentally in several ways:

1. High level 'interactive' or conversational 'languages' allow the engineer user to write and run his own operational programmes after a very short training (a few days). The computer is provided with a 'compiler' programme which translates the users 'high level' commands into the much less readily understood 'machine code'. There is, of course, a penalty, and this is manifest in the computing time taken for any mathematical operation. Luckily this is rarely serious for process control, as events take place very slowly.
2. Astonishingly rapid developments in electronic components and circuits have led to the small computer which, while having adequate speed and capacity

for control of an average process unit (refining column, reactor, etc.), cost (with auxiliaries) little or no more than the equivalent 'traditional' analog equipment. The basic reliability of such computers is, without doubt, vastly superior to that of earlier analog equipment, though this is not always obvious because of common reliance on less reliable 'field' mounted equipment.

3. Last but by no means least manufacturers have belatedly come to realize that the success of an on-line control system depends on the design and reliability of the input/output interface between the computer and the field mounted transmitters and final control elements. Much development has been necessary to achieve overall reliability in the process control computer 'package' available from several manufacturers today.

Although, in general, management are rightly sceptical of the real economic benefits to be expected from adopting advanced control techniques which require a computer for their implementation, there is one area of process control where the arguments are readily appreciated. This is in sequencing of the successive operations which together constitute a typical batch process (fill tank, weigh charge to reactor, heat up, evacuate, etc.). There is little difference between such control and a modern automatic washing machine except that in some instances the loop is closed by the feedback of information, (that a valve has indeed opened when commanded, for instance). Such operations can be readily initiated, either on a time lapse basis or on the basis of completion of a previous event, by a computer. They can just as easily be initiated by much simpler electro-mechanical or electronic 'sequencers' which can also be just as readily reprogrammed for a different sequence of events. Where the digital computer scores is that the comparison of the value of process variable or time at which an event is terminated or initiated can be made by the computer, rather than an external device which must be manually calibrated (temperature switch, totalizer, counter, etc.). Because of this a variety of 'recipes' can be preprogrammed for the same or different sequences and called up as required.

Apart from this advantage there is little to be said in favour of the digital computer for sequencing operations; yet it is in this area, because the advantages are readily understood that computers are becoming quite common. Fortunately the same computer can be used to advantage in implementing advanced process control techniques and this is of particular benefit in connection with the start up/run/shutdown nature of batch process. For instance, the integral saturation problem discussed in chapter 5 is easily overcome: non-standard algorithms, such as the use of an error squared proportional action in level control to give increasing gain as the error increases, can also readily be implemented.

The advantages which accrue from the use of a computer to control continuous processes are much greater than for the sequencing operation. Even if only standard single input/output loops are used the ability to change the measured value input, send the output to another control valve or 'cascade' it into the desired value of a second controller, confers enormous flexibility at the design stage, at start-up, and after. In fact, provided the necessary transmitters and control valves are provided, the total control scheme can be changed at any time; it is normally impracticable to make even a very minor change to an operating system with conventional analog controls.

The ability to implement some of the techniques outlined in this book (and eventually much more advanced techniques) for which the digital computer is, as has been seen, necessary, has yet to be seriously explored, but that will never be done until the principles of dynamic process behaviour are nearly as well understood by the chemical engineers who design the processes as are steady-state relationships. Control systems cannot continue to be added after the process design is complete like fairy lights on a Christmas tree. This book is written in the hope of bridging the gap which all too obviously exists between modern control theory and current practice in process design.

Appendix I Matrices

It is not within the scope of this book to deal in any depth with the subject of linear algebra; little has been assumed in the development of multivariable systems presented. It is necessary, in order to make use of the techniques discussed, to be able to calculate the 'determinant' and the reciprocal of a square matrix.

The determinant of a square matrix

Consider the equations of a system to be

$$a_1 x + b_1 y + c_1 z = 0 \qquad \text{(i)}$$
$$a_2 x + b_2 y + c_2 z = 0 \qquad \text{(ii)}$$
$$a_3 x + b_3 y + c_3 z = 0 \qquad \text{(iii)}$$

Combining (ii) and (iii) gives

$$\frac{x}{b_2 c_3 - b_3 c_2} = -\frac{y}{a_2 c_3 - a_3 c_2} = \frac{z}{a_2 b_3 - a_3 b_2}$$

Substitution in (i) gives

$$a_1(b_2 c_3 - b_3 c_2) - b_1(a_2 c_3 - a_3 c_2) + c_1(a_2 b_3 - a_3 b_2) = 0 \qquad \text{(iv)}$$

Rewriting the equations in matrix form

$$\begin{bmatrix} a_1 + b_1 + c_1 \\ a_2 + b_2 + c_2 \\ a_3 + b_3 + c_3 \end{bmatrix} \times \begin{bmatrix} x \\ y \\ z \end{bmatrix} = 0$$

It can be seen that the expression on the left of (iv) is given by the sum of terms obtained by multiplying each element of the first row by an expression obtained from the matrix remaining when the row and column containing that element are eliminated. This later expression is given by the difference of the products of the terms lying on the right diagonal and the product of the terms on the left diagonal. This expression is the determinant of the 'minor' or reduced matrix, while the full term (the left-hand side of (iv)) is the determinant of the whole matrix. Exactly the same result is obtained if any row or column is taken and each term multiplied by it's minor. Larger matrices are dealt with in the same manner, the calculation of the determinant of each minor of a fourth-order matrix being identical with the calculation of this third-order matrix. Because the set of equations is homogeneous in this case the determinant is zero, but in the general case it may have any scalar value.

The inverse of a square matrix

For convenience it is usual to drop the convention of $a, b, c, \ldots$ for the elements of the matrix and adopt instead the lower case of the matrix letter with two subscripted numbers, one denoting the position in a row, the other a position in a column. Thus for the matrix M each element is m_{ij} where i gives the row and j the column of the element.

The determinant of the 'minor' of any element of the matrix is shown as the 'cofactor' of that element. Replacing each element of the matrix by its cofactor multiplied by $(-1)^{i+j}$ and interchanging each row with the correspondingly numbered column (turning the matrix on it's side), we obtain the 'adjoint' matrix. The adjoint matrix of M has the property that

$$M \left[\frac{1}{|M|} \text{adj } M \right] = 1 \quad |M| \text{ is the determinant of } M$$

but by definition the inverse of a matrix must have the property

$$M . M^{-1} = 1$$

therefore

$$M^{-1} = \frac{1}{|M|} \text{adj } M$$

Appendix II System controllability

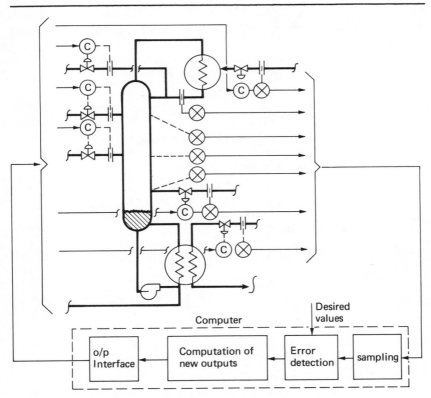

It will be readily appreciated that it is not always a simple matter to decide whether a system such as this is controllable by manipulation of the available inputs. Once the system relationships have been defined, however, and the transition matrix formed, we can establish controllability by scanning the system equation

$$\mathbf{x}(\tau) = e^{-[A]\tau}\mathbf{x}(0) + \int_0^\tau e^{[A](t-\tau)}[B]\mathbf{u}(t)\,dt$$

defining the states $x_1, x_2, \ldots, x_n$ as the difference between the present value and the desired value. For the system to be controllable, the state

$$\mathbf{x}(t) = 0$$

must be attainable in a finite time. Hence setting $\mathbf{x}(\tau) = 0$

$$e^{-[A]\tau}\mathbf{x}(0) + \int_0^\tau e^{[A](t-\tau)}[B]\mathbf{u}(t)\,dt = 0$$

or

$$\mathbf{x}(0) = -\int_0^\tau e^{[A]t}[B]\mathbf{u}(t)\,dt$$

As we already know, however, $e^{[A]t}$ can be expressed as the sum of powers of $[A]$ hence

$$e^{[A]t} = \sum_{i=0}^{(n-1)} \alpha_i(t)[A]^i$$

so that

$$\mathbf{x}(0) = -\sum_{i=0}^{(n-1)} [A]^i [B] \int_0^\tau \alpha_i(t)\mathbf{u}(t)\,dt$$

defining a new vector each element of which is

$$\beta_{j(t)} = -\int_0^\tau \sum_{i=0}^{(n-1)} \alpha_i(t)\mathbf{u}_j(t)\,dt$$

$\mathbf{x}(0)$ can be expressed in the form

$$\begin{bmatrix} x_1(0) \\ x_2(0) \\ \vdots \\ x_{(n-1)}(0) \end{bmatrix} [[B] + [A][B] + [A]^2[B] \cdots [A]^{(n-1)}[B]] \begin{bmatrix} \beta_1 \\ \beta_2 \\ \vdots \\ \beta_{(n-1)} \end{bmatrix}$$

Now each element $\beta_1, \beta_2, \ldots, \beta_{(n-1)}$ of the column vector is a multiple of $u_1, u_2, \ldots, u_{(n-1)}$ the controlling inputs and provided each column of the matrix

$$[B], [A][B], \ldots, [A]^{(n-1)}[B]$$

is independent we can be sure that *all* the states $x_1, x_2, \ldots, x_{(n-1)}$ can be influenced by at least one of the control inputs, which is to say that the system is fully controllable through its manipulable inputs.

Appendix III Observability of parameters

In the case of a multivariable non-independent system inability to 'observe' some system state may mean that we cannot control the system at all. However, just because of the system interactions, it is not always necessary to measure a system state directly; it can sometimes be inferred from measurement of other states. For instance, suppose that one of the states of a refining column system is the flowrate of vapour in an overhead line where no additional back pressure, such as a measuring orifice would impose, can be tolerated. We cannot measure flowrate, but we know that variations in this parameter affect the column temperatures and other parameters so that, given that the system is fully defined in the form of a transition matrix, it may be possible to deduce what change of overhead flow-rate is taking place from the behaviour of other system parameters.

Assuming that the measurable parameters of the system we wish to observe are combinations of the system states $x_1, x_2, \ldots, x_n$ then the vector of measured outputs $\mathbf{y}(t)$ is given by

$$\mathbf{y}(t) = [C]\mathbf{x}(t) \tag{A.1}$$

hence

$$\mathbf{y}(t) = [C]\mathbf{x}(t) = C\, e^{[A]t}\mathbf{x}(0) + C \int e^{[A](\tau-t)}[B]\mathbf{u}(t)\, dt \tag{A.2}$$

and the system can only be fully observed through the vector $\mathbf{y}(t)$ provided the matrix

$$[C]\, e^{[A]t}$$

fully defines the state space for the system. Expressing $e^{[A]t}$ as powers of the matrix $[A]$

$$[C]\, e^{[A]t} = [[C]^T, [A]^T[C]^T \cdots [A^{(n-1)}]^T[C]^T]$$

where $[A]^T$ or $[C]^T$ means the transpose of the matrix. The columns of the matrix are each formed by the product.

$$[A^i]^T[C] \tag{A.3}$$

where $[C]$ defines the state space spanned by the measured outputs and $[A]$

the state space for the system states. If there exists any direction in the system state space which is orthogonal to any direction definable in the space of [C] then the system cannot be observed through the vector. The system must be defined in the space defined by [C] or in some subspace of this. A subspace is one having dimensions which are all the same direction as *some* of the 'main' space. If this is so, it will be possible to generate a set of eigenvectors for the matrix $C e^{[A]t}$. This can only be true provided it is a square $n \times n$ matrix (n being the number of the system dimensions) which has no zero rows or columns.

Luenberger has shown that a satellite sub-system can be built from electrical, pneumatic, or hydraulic units which, when driven by the outputs of *an observable* system, will, after transients have died away, generate an anlog of the state which cannot be directly measured. Provided only that the transient behaviour can be made fast, in comparison with the process, then this analog can be used for control purposes exactly as any other measured variable.

Consider first a free system

$$\dot{x}(t) = [A]\dot{x}(t) \cdots x(t) = e^{[A]t}x(0) \tag{A.4}$$

Together with a satellite 'observer'

$$Z(t) = [D]Z(t) \cdots Z(t) = e^{[D]t}Z(0) \tag{A.5}$$

Provided both systems are time invariant, there will be some relationship between $Z(t)$ and $x(t)$: thus

$$Z(t) = [T]x(t) \tag{A.6}$$

where [T] is a unique matrix

then $\qquad [T]\dot{x}(t) = [D][T]x(t)$

and $\qquad \dot{x}(t) = [T]^{-1}[D][T]x(t)$

hence

$$[A] = [T]^{-1}[D][T] \tag{A.7}$$

also $\qquad Z(t) = e^{[D]t}Z(0)$

therefore $\qquad [T]x(t) = e^{[D]t}Z(0)$

and $\qquad x(t) = [T]^{-1} e^{[D]t}Z(0)$

Hence $\qquad e^{[A]t}x(0) = [T]^{-1} e^{[T][A][T]^{-1}t}Z(0)$

or $\qquad x(0) = e^{-[A]t}[T]^{-1} e^{[T]} e^{[A][T]t}Z(0)$

and $\qquad x(0) = [T]^{-1} e^{-[A]t} e^{[A]t} e^{[T]} e^{[T]^{-1}} Z(0)$

Hence

$$x(0) = e^{[T]^{-1}} Z(0) \quad (e \text{ is merely a constant}) \tag{A.8}$$

Similarly for a forced system

$$\dot{x}(t) = [A]x(t) + [B]u(t)$$

and
$$\dot{Z}(t) = [D]Z(t) + [G]u(t)$$

thus
$$[T]\dot{x}(t) = [D][T]x(t) + [G]u(t)$$

and
$$\dot{x}(t) = [T]^{-1}[D][T]x(t) + [T][G]u(t)$$

and thus
$$[D] = [T][A][T]^{-1}$$

and
$$[G] = [T][B] \qquad (A.9)$$

The meaning of this is that if two systems are related in a fixed manner (by $[T]$) their transition matrices will be similar and will thus have the same eigenvalues (a property of similar matrices). Thus the response of the two systems will be similar, that is, related by the matrix $[T]$.

If the 'observer' is stimulated by the states of the system so that for the system without inputs

$$\dot{x}(t) = [A]x(t)$$

$$\dot{Z}(t) = [D]Z(t) + [K][C]x(t)$$

then
$$Z(t) = [D]Z(t) + [K][C]x(t)$$

thus
$$[T]\dot{x}(t) = [T][A]x(t) = [D][T]x(t) + [K][C]x(t)$$

Hence
$$[T][A] = [D][T] + [K][C]$$

and
$$[A] = [T]^{-1}[D][T] + [T]^{-1}[K][C]$$

or
$$[D] = [T][A][T]^{-1} - [K][C][T]^{-1} \qquad (A.10)$$

The transition matrix of the observer is no longer $[T][A][T]^{-1}$ but is reduced by $[K][C][T]^{-1}$. If we now consider any error between the two system state vectors

$$\dot{Z}(t) - [T]\dot{x}(t) = [D]Z(t) + [K][C]x(t) - [T][A]x(t)$$
$$= [D]Z(t) + [T][A]x(t) - [D][T]x(t) - [T][A]x(t)$$

hence
$$\dot{Z}(t) - [T]\dot{x}(t) = [D][Z(t) - [T]x(t)]$$

and
$$Z(t) - [T]x(t) = e^{[D]t}[Z(0) - [T]x(0)]$$

Eventually $e^{[D]t}[Z(0) - [T]x(0)]$ will decay at a rate dependent on matrix $[D]$ so that

$$Z(t) = [T]x(t) \qquad (A.11)$$

after transients have decayed.

Since the system and the observer as constituted respond similarly, they will continue to do so when *both* are stimulated by the system inputs in similar fashion. Thus, if the 'observer' is forced by the system states modified by matrix $[T]$ and also by the system inputs similarly modified by $[T]$, it will respond in a similar fashion to the system itself and provide an estimate of the value of the state which cannot be measured directly. The observer satellite system can apparently be defined in almost any way provided only that it is correctly stimulated ($[T]$). This is so, *but* it must be remembered that the response of the observer differs from that of the system by reason of the transients

$$e^{[D]t}[Z(0) - [T]x(0)] \qquad (A.12)$$

In other words the responses are not identical until these transients die out. Since the rate at which this takes place is governed by the observer transition matrix $[D]$ it is necessary that the observer be 'fast' relative to the system to ensure that these transients are of very short duration. This, in turn, means that the observer must be built from components having little capacity and little inductance. The observer must not, of course, be unstable, and it can easily be shown that the factors or 'poles' of the observer characteristic equation must be different from any of those of the system. Hence a practical observer must:

1. Be stable.
2. Be 'fast'.
3. Have roots, factors, or poles different from those of the system.

Observers are in reality a special case of a filter. The observer is designed to give optimal response to the 'mode' or state variation required whilst at the same time rejecting as far as possible other 'modes' which can be considered as noise or unwanted signals. In other applications filters are used to reject unwanted random noise in measurement made of plant variables. Such random noise is normally of much higher frequency that any system 'mode': the observer must not be made so fast it tends to respond to such noise for, if it is, the 'observer' measurement will become unacceptably 'noisy' and thus useless. It can be seen, therefore, that the constitution of a practical observer is not arbitrary at all.

Bibliography

Symposium Series No. 32 Instn. Chem. Engrs. London (1969).
E. H. Bristol, 'A philosophy for single-loop controllers in a multiloop world', *Trans. Auto. Control I.E.E.*, The Foxboro Co. Massachusetts (1966).
H. H. Rosenbrock, 'Design of multivariable control systems using the inverse Nyquist array', *Proc. I.E.E.* (Control & Science) **116** (1966).

*F. G. Shinsky, *Process Control Systems,* McGraw-Hill (1967).
*W. L. Luyben, *Process Modelling, Simulation, and Control for Chemical Engineers,* McGraw-Hill (1973).
*H. H. Rosenbrock, *Mathematics of Dynamical Systems,* Nelson (1970).
R. E. Levy, A. S. Foss, and E. A. Grens II, 'Response modes of a binary distillation column', *I & E. C. Funadmentals,* 8, No. 4 (Nov. 1969).

* Recommended for further reading

Index

Adjoint, of a matrix, 145
Amplification (*see* Attenuation)
Analogue control, 90-91
    action, 77-83
        of distillation process, 91-96
Array, 92
Attenuation, 62
    (*see also* Magnitude ratio)
Auto-correlation, 137

Bandwidth, 79
Block diagrams, 35
Bode diagrams, 62-65
Boundary (*see* Interface)
Buffer storage, 8, 91

Caley-Hamilton theorem, 115
Capacity/capacitance, 11, 43-50
Characteristic equation, 24-28
Characteristic function, 43, 69-71, 78-83, 102
Closed loop, 35, 70
Closed system, 8
Complex:
    operator, 60
Continuous function, 98
Constraints, 12, 16
Controllability roots, 25, 39
    of system, 106, 145
Coordinates (*see* Dimensions)
Correlation function, 135-137
Coupling, 8-11, 37, 52, 109-112, 137-139
Cross-correlation, 137-139

Damping factor, 28, 53-54
Dead time, 67, 84
Decomposition of functions 112, 134
Decoupling (*see* Coupling)
Dependence, 12, 89, 104
Derivative control:
    action, 82-83
        in multivariable control, 118
        negative, 85

Determinant of matrix, 94, 100, 103, 144

Dimension of state space, 11, 92, 100, 130
Discrete function, 98
Disturbances, 115, 121
Divergent system, 10

Eigenvalue/eigenvector, 103
Elements of system, 6, 58
Environment of system, 6, 87, 98, 128
Equilibrium, 8
Error, 36, 77
Estimate (*see* Observer)
Exponential function, 17, 23

Feedback:
    control actions, 45-57
    in multivariable systems, 109-112
    effect on system behaviour, 35-41
Feedforward:
    control actions, 75, 78-82
    in multivariable systems, 118
Filtering of signals, 112
Forcing function, 29
Fourier transform, 32
Freedom, degrees of, 12-15
Frequency response, 60
Functions of matrices, 115

Gain:
    open loop, 93
    process, 38
    (*see also* Attenuation and Bandwidth)

Hold-up, 8
    (*see also* Buffer storage)

Identification of process, 128
Independance (*see* Dependance)
Inertia:
    in control mechanisms, 55-57
    in process, 50-55
Inherent stability (*see* self-regulation)
Initial conditions, 26
Instability, 18, 50
Integral control action, 75, 78-82
    in multivariable systems, 118

Integral transforms (*see* Fourier and Laplace)
Interaction, 95, 99
   (*see also* Coupling)
Interface, 7, 58
Inverse:
   of matrices, 143
   Nyquist arrays, 110
   response, 85, 109

Laplace transform, 33

Magnitude ratio, 34, 40, 60–69
Matrix, 94, 143
   (*see also* Array)
Models:
   dynamic state, 120–127
   in multivariable control, 95
   steady state, 98
   of systems, 87
Multivariable systems, 87–134

Natural frequency, 30, 70
Nyquist diagrams, 68

Observability, 147
Observer, 148
Off-set (*see* Error)
Open loop, 70
Open system, 8
Operating point, 14
Optimality, 87, 129
Orthogonality, 92
Oscillation, 20
Overshoot, 78

Phase angle, 34, 56, 60–76
Phase gain diagrams (*see* Bode and Nyquist)
Proportional control action, 46, 53, 77
Pulse testing, 133

Random stimuli, 137
Regulation (*see* self-regulation)
Relating function, 34, 59, 76
Relative gain array, 95
Reset, 77
   (*see also* Integral action)
Resistance, 43–57
Resolution, level of, 7
Resonance, 30
Response, 6, 29, 36, 60
Roots:
   of characteristic equation, 25
   of multivariable systems, 106
   of polynomial functions, 38
Root locus, 41, 69–76

Sample system (*see* Sampling period)
Sample theorem, 113
Sampling period, 112–116
Saturation of integral action, 79
Scalar, 102
Scale of matrices, 133, 144
Self-regulation, 8–9, 109, 130
Sensitivity (*see* Magnitude ratio)
Sinusoidal forcing, 30, 34, 60, 134
State (*see* System state)
Steady state, 10, 42
Step function, 32
Stimuli, 6, 29, 36, 60
Sub-systems, 87, 97, 107
System state, 8, 87, 129

Testing (*see* Sinusoidal forcing and Pulse testing)
Time constants, 43
Transfer functions, 30, 42, 67
Transform (*see* Fourier and Laplace)
Transient behaviour, 12, 18, 79, 135
Transition matrices, 101, 149
Transport delay (*see* Dead time)

Vectors, 93